THE SHADOW THAT WEIGHS

A Layperson's Journey Into How Physicists Measure Things That Have No Mass But Bend Reality

Adrian Kesterholm

TABLE OF CONTENTS

CHAPTER ONE

The Weight of Nothing

In the autumn of 1933, a Swiss astronomer with a reputation for being difficult stood at a telescope in Southern California and did something that would take the rest of the twentieth century to understand. Fritz Zwicky was not, by most accounts, an easy man to work with. Colleagues at Caltech found him abrasive. He called people he disagreed with "spherical bastards," on the theory that they were bastards no matter which way you looked at them. But on this particular night, working at the Mount Wilson Observatory, Zwicky was not thinking about his colleagues. He was thinking about a cluster of galaxies more than three hundred million light years from Earth, a swarm of roughly a thousand galaxies known as the Coma Cluster, and he was doing something that sounds simple but was, at the time, close to unprecedented. He was weighing it.

You cannot put a galaxy cluster on a scale. What Zwicky did instead was use two independent methods to estimate the mass of the Coma Cluster and see if they agreed. The first method was the obvious one: count the galaxies, estimate how much they glow, and use that light output to infer how much matter must be sitting there producing it. Stars shine because they burn, and the brighter a galaxy is, the more stars it presumably contains, and the more mass. This is straightforward astrophysics, the kind of calculation that had been refined for decades.

The second method was cleverer, and it relied on an idea borrowed from an unlikely source: the way molecules in a gas move. In 1933, Zwicky applied something called the virial theorem, a piece of physics normally used to relate the temperature of a gas to the speed of its molecules, to an entire cluster of galaxies. The insight was this: if you know how fast the galaxies in a cluster are moving relative to each other, and you assume the cluster is gravitationally bound (meaning it is not flying apart), you can calculate how much mass must be present to hold everything together with that much gravity. Fast-moving galaxies in a stable cluster require more mass to keep them from escaping. It is the same physics that tells you how fast a moon must orbit a planet to stay in orbit rather than drifting away or crashing down.

Zwicky measured the velocities of the galaxies in the Coma Cluster using the Doppler shift of their light, the same principle that changes the pitch of an ambulance siren as it passes you. Redder light meant a galaxy moving away faster; the spread of velocities told him how vigorously the whole cluster was churning. Then he ran the numbers.

They did not agree. Not even close. The mass needed to explain how fast the galaxies were moving, the mass required to keep the Coma Cluster from flying apart, was roughly four hundred times greater than the mass he could account for by counting up the visible light. Zwicky published this finding in a Swiss journal in 1933, using the German phrase "dunkle Materie," dark matter, to describe whatever was making up the difference. He was not being poetic. He meant it descriptively: matter that was dark, meaning it emitted no detectable light, but whose gravitational fingerprints were unmistakable.

For decades, almost nobody followed up on this. Zwicky's estimate was rough, his personality made him easy to dismiss, and the astronomical community of the 1930s had bigger and more tractable problems. The idea of dark matter sat in the scientific literature the way an unclaimed lottery ticket sits in a coat pocket: real, valuable, and entirely ignored. It would take nearly forty years, and a very different kind of scientist working a very different kind of problem, before anyone took the missing mass seriously again. That scientist, an astronomer named Vera Rubin, appears in the next chapter of this book, and what she found did not just confirm Zwicky's hunch. It transformed a curiosity into arguably the greatest open question in physics.

But before we get to Rubin, it is worth sitting with the strangeness of what Zwicky had actually stumbled onto, because it sets up everything else in this book. He had found evidence, using nothing but the motion of galaxies and a piece of nineteenth century statistical mechanics, that the universe contained far more mass than anyone could see. Not mass hidden behind something, the way the far side of the moon is hidden. Mass that produced no light of any kind, at any wavelength, and yet exerted a gravitational pull four hundred times stronger than what the visible galaxies could account for.

This book is about that gap, and about three other gaps like it, all of which share a strange family resemblance: things that appear to have no mass, or at least nothing you could hold, weigh, or photograph directly, and yet demonstrably shape the structure, expansion, and behavior of the entire universe. Dark matter is the first of these. Dark energy, the mysterious force accelerating the expansion of the cosmos, is the second. Gravitational waves, ripples in the fabric of spacetime itself that carry no mass but carry enormous energy across billions of light years, are the third. And the restless, foaming energy of empty space, what physicists call the quantum vacuum, is the fourth. Each is, in its own way, a shadow that weighs something. Each was detected not by seeing it directly but by observing its effects on things we can see. And each, when you actually dig into the story of its discovery, turns out to be one of the great detective stories of modern science.

Here is the promise of this book: by the time you finish it, you will understand, in real and specific terms, how physicists went about measuring things that cannot be seen, touched, or held. Not in vague poetic language about "the mysteries of the cosmos," but in concrete terms: what instruments were built, what calculations were run, what mistakes were made and corrected, and what the current state of knowledge actually is as of this writing. You will understand why roughly eighty five percent of all the matter in the universe is invisible to every telescope ever built. You will understand why the universe is not just expanding but expanding faster every year, pushed by something scientists frankly do not understand yet. You will understand how, in 2015, humanity detected a ripple in spacetime itself, created by two black holes colliding more than a billion years ago, using an instrument sensitive enough to notice a change in distance one ten-thousandth the width of a proton. And you will understand why even empty space, the vacuum between galaxies, between atoms, between the particles that make up your own body, is not empty at all, but is instead a roiling sea of energy that never quite reaches zero.

None of this requires a physics degree to follow. It requires curiosity and a willingness to sit with a few equations long enough to see what they are actually saying, which is different from being intimidated by them. Where equations appear in this book, and they will, they are the load-bearing kind: the ones that changed how physicists thought, presented with enough explanation that you will understand not just what they say but why they mattered. This is not a textbook. It is closer to a series of detective stories, each one built around a genuine historical mystery, each one resolved (or, in some cases, still unresolved) by real scientists using real instruments, working with the same combination of stubbornness, cleverness, and occasional dumb luck that characterizes most of the history of science.

Let us return, for a moment, to Zwicky's number: four hundred times more mass than expected. It is worth pausing on how audacious a claim this was in 1933. Newton's laws of gravity, verified over two and a half centuries, describe with extraordinary precision how planets orbit the sun, how the moon orbits the earth, how a thrown ball arcs through the air. Nobody in 1933 doubted that gravity, applied correctly, should predict how a cluster of galaxies behaves. So when Zwicky's calculation produced a wild mismatch, there were, broadly, three possible explanations, and the history of the last ninety years of physics has largely been the story of investigating each of them in turn.

The first possibility was that Zwicky had simply made an error, either in his velocity measurements or in his estimate of the cluster's light output. This is always the first and most boring explanation for any anomaly in science, and it is usually the correct one. Most reported anomalies in physics turn out to be measurement error, and for decades this was the safest bet on Zwicky's missing mass too. Telescopes in 1933 were primitive by modern standards. The Doppler shift measurements were imprecise. It was entirely reasonable to suspect the whole thing would dissolve under better data.

The second possibility was that Newton's laws of gravity, so reliable at the scale of the solar system, simply broke down at the much larger scale of galaxy clusters. This is not as crazy as it sounds. Physics is full of examples where a law that works perfectly well in one regime fails in another. Newtonian mechanics itself gives way to Einstein's relativity at very high speeds and very strong gravitational fields. Perhaps, some physicists reasoned, gravity itself behaves differently across the vast distances separating galaxies within a cluster, and no missing matter is needed at all, just a correction to the law of gravity itself. This idea has never entirely gone away. A minority of physicists still pursue modified gravity theories, and later in this book we will look honestly at where that research stands and why the mainstream of the field has moved in a different direction.

The third possibility, the one that eventually won out, was that Zwicky was right: there really is more mass out there than we can see, made of something that does not emit, reflect, or absorb light in any way we have yet detected, but that nonetheless has mass and therefore exerts gravity, the same as anything else in the universe. This is the possibility this book takes most seriously, not because it is more comforting or more elegant, but because it is the explanation that decades of subsequent, far more precise observations have overwhelmingly supported. When we get to Vera Rubin's work in the next chapter, and then to gravitational lensing, and then to the cosmic microwave background, and then to the Bullet Cluster (one of the single most persuasive pieces of evidence in all of modern astrophysics), you will see why the scientific consensus settled where it did, and you will be equipped to evaluate that evidence yourself rather than simply take it on faith.

It is worth saying clearly, at the outset of this book, that dark matter has never been directly detected. As of this writing, no experiment on Earth has captured a dark matter particle, examined it, and confirmed what it is. This might sound like a devastating admission for a book that is about to spend several chapters discussing dark matter as though it were established fact. It is not. The gravitational evidence for dark matter's existence, accumulated across nine decades and dozens of independent lines of investigation using entirely different methods, is about as solid as anything in astrophysics. What remains genuinely open, and what chapter four and five of this book explore in real detail, is the question of what dark matter actually is at the particle level, and that is a very different kind of uncertainty than doubting whether it exists at all. Consider an analogy: you can be completely certain that wind exists, because you can watch it bend trees and feel it on your skin, without being able to see a single air molecule. Dark matter's existence is inferred the same way, through its effects, even though the particle or particles making it up remain, for now, unidentified.

The same basic pattern, effect observed, cause inferred, mechanism still being worked out, recurs throughout this book. Dark energy was discovered not by detecting some energy field directly but by noticing that distant supernovae were dimmer than they should have been, implying the universe's expansion is accelerating rather than slowing down, a finding so unexpected that when the two research teams first saw it in their data in 1998, several scientists on both teams initially assumed they had made a mistake. Gravitational waves were predicted by Einstein in 1916 as a mathematical consequence of general relativity, and Einstein himself later doubted they would ever be strong enough to detect, yet in 2015 a pair of underground detectors in Louisiana and Washington State picked up the unmistakable signature of two black holes spiraling into each other and merging, over a billion light years away. And the quantum vacuum, seemingly a contradiction in terms (how can a vacuum, defined as the absence of anything, do anything at all), turns out to seethe with energy that has been measured in laboratories through an effect named after the Dutch physicist who predicted it, Hendrik Casimir.

Each of these four phenomena, dark matter, dark energy, gravitational waves, and the quantum vacuum, gets its own extended treatment in the chapters ahead, generally two to four chapters apiece, moving from the historical discovery through the current experimental frontier. But they are not four unrelated topics stitched together for convenience. They share something deeper, which is the actual subject of this book: each one represents an occasion where physicists were forced to conclude that something without a directly observable form, something you could reasonably have called "nothing" from a naive standpoint, in fact carries real physical consequence. Real weight, in the broadest sense of the word. Real influence on how galaxies form, how the universe expands, how black holes collide, and how energy itself behaves at the smallest imaginable scales.

It is worth being explicit, at the very outset, about why these four phenomena belong together in a single book rather than four separate ones, because the connection is not merely thematic convenience. Each was discovered using a broadly similar epistemological strategy: physicists noticed that some well established, thoroughly tested theory, Newtonian gravity, general relativity, quantum field theory, made a prediction that did not match what could actually be observed, and rather than discarding the theory itself, which had already proven its worth across an enormous range of other predictions, they asked what additional, previously unaccounted ingredient would need to be present to reconcile prediction with observation. This is a conservative, methodologically disciplined way of responding to an anomaly, and it stands in deliberate contrast to a more radical alternative: throwing out the underlying theory altogether and starting from scratch. Both approaches have their place in the history of physics, and this book will touch on serious, credentialed attempts to pursue the more radical path, particularly modified gravity theories addressed in chapter thirteen. But the conservative path, preserve the theory, identify the missing ingredient, has been the dominant and most productive strategy across all four investigations this book follows, and understanding why it has worked so consistently well is one of the quieter, more durable lessons this book hopes to leave you with.

There is a reason this matters beyond intellectual curiosity, and it is worth stating plainly before we go further. Roughly ninety five percent of the universe, by most current estimates, is made up of dark matter and dark energy combined. The ordinary matter that makes up you, this book, the earth, the sun, every star visible to the naked eye, accounts for something like five percent of the total mass and energy content of the cosmos. We are, in the most literal sense, a minority phenomenon in our own universe, adrift in a sea of stuff we still do not fully understand. That is either an unsettling thought or an exhilarating one, and this book is written on the bet that, once you understand how physicists actually arrived at that number, and how they are working to pin down the rest, you will find it closer to exhilarating.

One more thing before we move forward. Popular science writing about physics has a tendency to reach for grandeur: cosmic mystery, the fabric of the universe, the secrets of creation. Some of that language will show up in this book too, because some of these discoveries genuinely warrant it. But underneath the grandeur, what you are actually going to read about is a long sequence of extremely specific, extremely human achievements: an astronomer squinting at photographic plates in the desert night, engineers building a detector sensitive enough to notice a vibration a thousand times smaller than a proton, physicists arguing over dinner about whose calculation was wrong, technicians a mile underground in an old gold mine in South Dakota trying to keep cosmic rays from contaminating a tank of liquid xenon. Science, at the level where it actually happens, looks a great deal less like poetry and a great deal more like exacting, frustrating, occasionally thrilling craftsmanship. That is the story this book is actually telling, and it starts, properly, with a woman at a telescope in the 1960s who could not stop looking at a puzzling printout, and who changed astronomy forever because she refused to look away.

CHAPTER TWO

Zwicky's Missing Mass

Here is a question worth sitting with for a moment: how do you weigh something you cannot touch, sitting three hundred and twenty million light years away?

It sounds like a trick question, or the setup to a joke about physicists. It is neither. It is the actual problem Fritz Zwicky solved in 1933, and the method he used is elegant enough that it is worth walking through slowly, step by step, because the same basic logic reappears again and again throughout this book, in gravitational lensing, in the study of the cosmic microwave background, and in the Bullet Cluster evidence discussed later in this book. Once you understand how Zwicky weighed the Coma Cluster, you have the conceptual toolkit for understanding how astronomers weigh almost anything too large or too distant to put on a scale.

Start with what you can actually observe. A galaxy cluster like Coma is a collection of individual galaxies, each one containing hundreds of billions of stars, all bound together by their mutual gravity, orbiting a common center of mass the way planets orbit a sun, except messier and far more chaotic. From Earth, using a telescope, you can measure two things about each galaxy in the cluster with reasonable precision. First, you can measure how bright it appears, and if you know the distance to the cluster (which astronomers can determine through a separate set of techniques involving standard candles, discussed further in chapter six), you can convert that apparent brightness into an estimate of how much light the galaxy is actually producing. Second, using a technique called spectroscopy, you can measure the galaxy's velocity relative to Earth by looking at how much its light has shifted toward the red or blue end of the spectrum, the same Doppler effect that changes the pitch of a passing train's whistle.

Neither of these measurements tells you mass directly. Light output tells you about the number and type of stars in a galaxy, and there is a reasonably reliable relationship (astronomers call it the mass to light ratio) between how much light a collection of stars produces and how much mass those stars represent. This was Zwicky's first method: sum up the light from every galaxy in the Coma Cluster, apply a standard mass to light ratio, and arrive at a total mass estimate. Call this the luminous mass.

The velocity measurements are where things get more interesting, and where the second, independent method comes in. Zwicky measured the velocities of dozens of galaxies within the Coma Cluster and found that they were moving remarkably fast relative to one another, with a velocity spread (astronomers call this velocity dispersion) of around a thousand kilometers per second. That is fast. For context, the Earth orbits the Sun at about thirty kilometers per second. The galaxies in the Coma Cluster were churning around their common center of mass more than thirty times faster than our planet moves around its star.

Here is the crucial physical insight, and it is worth stating precisely because everything else in the chapter depends on it. A cluster of galaxies moving at that velocity, held together by nothing but the gravity generated by the visible stars within those galaxies, should not be stable. The gravity from the visible matter simply is not strong enough to keep galaxies moving that fast from flying off into intergalactic space. If you calculate how much total mass would be required to generate enough gravitational pull to hold a cluster together at those velocities, using the same virial theorem physicists use to relate the temperature of a gas to the speed of its molecules, you get a number. Zwicky calculated that number for Coma, and it came out to roughly four hundred times greater than the luminous mass estimate.

Put differently: either the Coma Cluster was in the process of flying apart, a wildly unstable structure that happened to be caught, by pure chance, at a moment in cosmic history when we could observe it, or there was several hundred times more mass present than all the visible stars and galaxies could account for, mass that was providing the additional gravitational glue holding the whole system together.

Zwicky, characteristically unbothered by how radical this sounded, went with the second explanation, and he coined the term that has stuck for ninety years: dunkle Materie. Dark matter.

It would be satisfying to report that the astronomical community responded to this with immediate scientific excitement and a wave of follow-up studies. It did not. Several factors conspired against Zwicky's finding gaining traction in the 1930s. His personality did not help. He was known for open contempt toward colleagues he considered slow or unoriginal, and this made him plenty of enemies within the small, insular world of American astronomy. There were also legitimate scientific reasons for skepticism. The virial theorem calculation depended on assumptions, including the assumption that the Coma Cluster was in a stable, gravitationally bound state (rather than a transient, dispersing structure), and on 1930s-era measurements of galaxy velocities and distances that carried substantial uncertainty. A four hundred-fold discrepancy is enormous, and enormous discrepancies invite enormous skepticism, which is exactly the correct scientific instinct.

There is a version of this story where Zwicky's insight simply gets forgotten, filed away as an eccentric Swiss astronomer's miscalculation, and dark matter never becomes a serious subject of study. That almost happened. For nearly four decades, the missing mass problem sat largely dormant in the scientific literature, occasionally mentioned, rarely investigated with any urgency. A handful of astronomers in the 1930s and 1940s, including the Dutch astronomer Jan Oort, found hints of similar discrepancies when studying the motion of stars within our own galaxy, and a few theorists kept the idea alive as a topic worth eventual investigation. But it was not until the 1970s, when an entirely different line of research, focused not on galaxy clusters but on the internal rotation of individual spiral galaxies, produced evidence so clean and so hard to dismiss that the astronomical community could no longer look away, that dark matter moved from curiosity to central mystery. That story belongs to the next chapter, and to Vera Rubin.

But it is worth pausing here, at the end of Zwicky's story, to draw out a broader point about how science actually progresses, because it is a pattern that recurs throughout the rest of this book. Scientific ideas do not typically succeed the moment they are first proposed, no matter how correct they later turn out to be. They succeed when enough independent lines of evidence, gathered by enough different researchers using enough different methods, converge on the same conclusion. Zwicky's 1933 calculation was, in hindsight, essentially correct. The Coma Cluster really does contain roughly the amount of extra mass he calculated, a conclusion confirmed by multiple independent modern techniques, including gravitational lensing measurements taken decades later that we will examine in chapter four. But being correct in 1933 was not, by itself, enough to convince the field. Zwicky needed forty years of additional evidence, gathered by other scientists, using other methods, before dark matter became something the astronomical community took seriously as a central problem rather than an eccentric footnote.

This matters for how you should think about the other three mysteries in this book too. Dark energy, discovered in 1998, gained rapid acceptance partly because two independent research teams, working separately and in friendly competition with each other, arrived at the same startling conclusion using the same basic method (studying distant supernovae) but analyzing different data sets. Gravitational waves, predicted by Einstein in 1916, waited nearly a century for direct detection, not because physicists doubted Einstein's math, but because building an instrument sensitive enough to actually measure them required advances in laser technology, vacuum engineering, and vibration isolation that simply did not exist until the early twenty first century. Good science is, in a very real sense, patient. It has to be. The universe does not organize itself for our convenience, and the tools required to measure its stranger corners often have to be invented from scratch.

There is one more piece of Zwicky's story worth including here, because it illustrates something important about the man and about how scientific reputations get made and unmade. In addition to his work on dark matter, Zwicky made a series of other genuinely important contributions to astrophysics during his long career at Caltech. Working with the astronomer Walter Baade in the 1930s, he was among the first to propose that supernovae (the explosive deaths of massive stars) could produce neutron stars, incredibly dense stellar remnants where an entire star's worth of mass gets compressed into a sphere only a few miles across. This idea, radical at the time, was later confirmed and is now standard astrophysics. Zwicky also compiled catalogs of galaxies and galaxy clusters that remain useful reference works. And yet, for decades, he was remembered in the field more for his abrasive personality than for the prescience of his science. It is a useful reminder that being right and being believed are not the same accomplishment, and that the second often takes considerably longer than the first.

By the time Zwicky died in 1974, the tide had just begun to turn. Vera Rubin's early rotation curve work was underway, and though Zwicky did not live to see dark matter become one of the central research programs of modern physics, the seeds of that transformation were already germinating in observatories in Arizona and Washington, D.C., where a very different kind of astronomer was staring at printouts of her own, equally puzzled by what she was seeing, and equally unwilling to let the puzzle go unexplained.

It is worth taking a moment to think about scale, because the numbers involved in this chapter are easy to read past without absorbing. The Coma Cluster contains, by modern estimates, over a thousand identified galaxies, and possibly closer to ten thousand when fainter members are included. Each of those galaxies contains, on average, hundreds of billions of stars. The cluster itself spans roughly twenty million light years across, meaning light, traveling at 299,792 kilometers per second, the fastest anything can move in the universe, takes twenty million years just to cross from one side of the cluster to the other. And Zwicky's calculation implied that the visible galaxies, all one thousand to ten thousand of them, all those hundreds of billions of stars, accounted for perhaps one four-hundredth of the cluster's actual gravitational mass. Modern estimates have revised that ratio somewhat (current figures suggest dark matter outweighs ordinary matter in galaxy clusters by a factor closer to five or six to one, once you account for hot intracluster gas that Zwicky's era could not observe, since gas does not necessarily contribute the way stars do), but the essential shock of the finding survives intact. The vast majority of the mass in the largest gravitationally bound structures in the universe is, and remains, invisible.

Sit with that for a moment before we move to the next chapter, because it is the foundation everything else in this book rests on. The universe is not what it looks like. Look up on a clear night, away from city lights, and the stars scattered across the sky represent a small minority of what is actually out there, gravitationally speaking. The rest is dark: not empty, not absent, but dark in the specific sense Zwicky meant when he coined the phrase. Present. Massive. Gravitationally active. And, as of this writing, still not directly detected, ninety three years after a difficult Swiss astronomer first did the math and refused to explain away what it told him.

There is also a methodological lesson buried in the Coma Cluster story that is worth drawing out explicitly, because it shapes how the rest of this book approaches evidence. Zwicky's calculation depended on two separate, independently measured quantities: how bright the cluster's galaxies appeared, and how fast those galaxies were moving. These measurements came from entirely different physical processes. Brightness depends on nuclear fusion inside stars, the same process that powers the Sun, converting hydrogen into helium and releasing light as a byproduct. Velocity, measured through the Doppler shift, depends on nothing more than the wave nature of light itself and the relative motion between source and observer, a phenomenon confirmed in laboratories on Earth long before it was ever applied to galaxies. The fact that two measurements derived from such different physics, one rooted in nuclear astrophysics, the other in the wave mechanics of light, could be combined mathematically to expose a genuine gap in our accounting of the universe's mass, is precisely the kind of independent convergence that makes a scientific claim credible. It is not proof by itself. But it is the opposite of a coincidence, and throughout the rest of this book you will see the same pattern recur: physicists reaching for two or more genuinely unrelated methods to check whether they agree, and treating agreement between independent methods as far stronger evidence than any single measurement, however precise, could ever provide on its own.

Consider, for a moment, how differently the history of physics might have unfolded had Zwicky's 1933 result simply vanished from the record instead of sitting, dormant but recorded, in the scientific literature for four decades. Ideas that never get published do not get revived, because there is nothing for a later researcher to stumble across, cite, and build upon. Zwicky's insistence on publishing his result, unfashionable and difficult to explain as it was, meant that when Vera Rubin and others began finding their own independent evidence for missing mass in the 1970s, there was already a name for the phenomenon, already a rough order-of-magnitude estimate to compare against, and already a published precedent establishing that at least one prior scientist had looked at similar data and drawn the same startling conclusion. Scientific priority, the credit for being first, gets discussed constantly in the history of science, sometimes to an unhealthy degree. But there is a genuinely important reason it matters beyond ego or reputation: a published, dated, peer-reviewed record is what allows later scientists to build cumulatively on earlier work, rather than each generation rediscovering the same puzzles from scratch. Zwicky's 1933 paper, largely ignored for a generation, was nonetheless doing quiet work the entire time, sitting in library shelves and journal archives, waiting for the field to catch up to what one difficult, brilliant astronomer had already worked out with a telescope, a slide rule, and a stubborn refusal to explain away an inconvenient number.

CHAPTER THREE

Vera Rubin and the Spinning Galaxies

It was 1965, and Vera Rubin had just been hired for a part-time position at the Department of Terrestrial Magnetism at the Carnegie Institution in Washington, a job that, on paper, had nothing to do with dark matter, galaxy clusters, or any of the questions Fritz Zwicky had raised three decades earlier and that the field had largely stopped asking. Rubin, a mother of four with a PhD from Georgetown, had already spent a decade working on other people's research problems, unable to find a full research position that took her scientific ambitions seriously. Astronomy in the mid-twentieth century was not a welcoming field for women. Rubin had been advised, more than once, to pick a research topic modest enough that nobody would feel threatened by it.

At Carnegie, she met a physicist named Kent Ford, who had built something genuinely useful: an image tube spectrograph, a device that dramatically increased the sensitivity of spectroscopic measurements, cutting the exposure time needed to capture a usable spectrum down to a fraction of what older equipment required. Rubin recognized immediately what this instrument made possible. She could use it to measure, with far greater precision than earlier astronomers had managed, exactly how fast stars and gas clouds were moving as they orbited within a single galaxy, rather than how fast entire galaxies moved within a cluster, which had been Zwicky's approach.

She and Ford chose to start with the Andromeda Galaxy, the nearest large spiral galaxy to our own Milky Way, close enough and bright enough to study in detail. Their plan was straightforward, at least in concept: measure the velocity of stars and glowing hydrogen gas at many different distances from Andromeda's center, and plot how orbital velocity changes with distance. This plot is called a rotation curve, and in 1965, astronomers had a very clear expectation for what it should look like.

The expectation came directly from Newtonian physics, the same physics that predicts how planets orbit the sun. In our solar system, Mercury, close to the sun, orbits fast, at nearly fifty kilometers per second. Neptune, far more distant, orbits slowly, at just over five kilometers per second. This falloff in orbital speed with distance, technically called Keplerian motion after the astronomer who first described planetary orbits mathematically, happens because most of the mass in the solar system, the sun itself, is concentrated at the center. Objects farther from that concentrated mass feel a weaker gravitational pull and settle into slower orbits.

Astronomers in the 1960s assumed galaxies would work the same way, just at a different scale. Most of a galaxy's visible mass, its stars, is concentrated toward the galactic center, in the bright central bulge you can see in photographs of spiral galaxies. So the expectation was that stars near the center of Andromeda would orbit fast, and stars far out toward the edges of the galactic disk, in the sparser, dimmer regions, would orbit progressively slower, mirroring the pattern of planets around the sun.

Rubin and Ford began taking measurements at Lowell Observatory in Flagstaff, Arizona, and at Kitt Peak National Observatory, working through long nights, trading off between operating the telescope and processing photographic plates as they developed. Rubin later described those nights in vivid detail: alternating turns with Ford between the two tasks, eating ice cream cones while waiting for exposures to develop, gradually building up a data set point by point. By the end of the very first night of observations, something was clearly wrong, or at least clearly not matching expectations. "The surprises came very quickly," Rubin recalled decades later. "By the end of the first night, we were puzzled by the shape of the rotation curve."

The puzzle was this: the rotation curve was not falling off with distance the way Keplerian motion predicted. Instead, it was flat. Stars far out at the edge of Andromeda's visible disk were orbiting at essentially the same speed as stars much closer to the center. This should not have been possible if the vast majority of the galaxy's mass were concentrated in the visible central bulge, the way the sun's mass dominates the solar system. A flat rotation curve implied something quite different: that mass was distributed throughout the galaxy, extending well beyond the visible disk of stars, roughly uniformly enough that the gravitational pull, and therefore the orbital speed, stayed remarkably constant regardless of distance from the center.

Rubin and Ford published their initial Andromeda rotation curve findings in 1970, in a paper that extended their optical measurements out to 110 arc minutes from the galaxy's center, a huge span that reached well past the visible starlight into the fainter outskirts of the galaxy. Their results lined up with earlier radio observations made by the astronomer Morton Roberts, who had used a completely different technique, radio measurements of neutral hydrogen gas rather than optical measurements of starlight, and arrived at a similar conclusion. Two independent methods, converging on the same strange answer, is exactly the kind of evidence that makes scientists sit up and take notice, and it was around this same period, 1970, that other researchers began independently publishing papers arguing that galaxies required additional unseen mass to explain their observed rotation.

Rubin did not stop with Andromeda. Over the following years, working with Ford and later with the astronomer Norbert Thonnard, she systematically extended the same technique to dozens of additional spiral galaxies, and the pattern held with remarkable consistency. Galaxy after galaxy showed flat, or nearly flat, rotation curves rather than the Keplerian falloff Newtonian physics predicted based on visible mass alone. This was the critical difference between Rubin's work and Zwicky's earlier finding. Zwicky had found one anomalous cluster and proposed a bold explanation for it. Rubin, working systematically across an entire sample of galaxies with careful, repeatable methodology, found the same anomaly everywhere she looked. It was not a fluke, not a single misbehaving cluster that might be explained away by unusual local circumstances. It was a universal property of spiral galaxies.

Let us walk through the actual physics here, because it rewards close attention and it is not complicated once laid out clearly. For an object orbiting a central mass under gravity alone, in a circular orbit, physics gives us a clean relationship between orbital velocity and distance. Set the gravitational force equal to the centripetal force required to maintain a circular orbit, and you get v equals the square root of GM divided by r, where v is orbital velocity, G is the gravitational constant, M is the mass enclosed within the orbit, and r is the orbital radius. If M stays constant as r increases, which is approximately true once you are far enough out that additional mass is negligible, velocity should decrease as the square root of distance increases. That is Keplerian falloff, and it is exactly what we observe in the solar system.

But if velocity stays flat as r increases, as Rubin observed, then the equation tells you something must be changing to compensate. Specifically, M, the mass enclosed within the orbit, must be increasing roughly in proportion to r, growing larger the farther out you look, rather than staying constant. This is the mathematical signature of an extended, roughly spherical halo of mass surrounding each galaxy, a halo whose total mass grows with radius rather than being concentrated at the center the way visible starlight is. Since this additional mass produces no detectable light of any kind, at any wavelength astronomers have checked, it earned the same name Zwicky had proposed decades earlier: dark matter.

The implications, once astronomers absorbed them, were staggering. If Rubin's flat rotation curves held across galaxies generally, and they did, then most of the mass in a typical spiral galaxy, including our own Milky Way, was not the visible stars, gas, and dust that show up in photographs. It was an invisible halo, extending well beyond the edge of the visible disk, outweighing the visible matter by a wide margin. Modern estimates place the ratio at roughly five or six to one, meaning for every unit of ordinary visible matter in a typical galaxy, there are roughly five to six units of dark matter, arranged in a rough sphere that envelops the visible disk the way a fog envelops a lit streetlamp.

Rubin's work succeeded in convincing the field where Zwicky's had not, for a combination of reasons worth naming explicitly. First, the sheer volume and consistency of her data mattered enormously. This was not one anomalous measurement; it was dozens of galaxies, all showing the same pattern, gathered with careful, repeatable methodology using state of the art instrumentation. Second, the timing mattered. By the 1970s, theoretical physicists including Jeremiah Ostriker and Jim Peebles at Princeton had already begun running computer simulations suggesting that spiral galaxies needed additional unseen mass simply to remain structurally stable over cosmic timescales, without the disk flying apart or collapsing. Rubin's observations arrived at a moment when theory was already primed to take the idea seriously. Third, and this should not be understated, Rubin herself was a scrupulously careful scientist with a reputation, hard won over a difficult career, for meticulous, trustworthy work. When she reported a result, the field had learned to trust it.

It is worth pausing to note what Rubin's discovery did and did not establish. It provided overwhelming evidence that galaxies contain far more mass than their visible stars can account for, mass that behaves gravitationally exactly as ordinary matter would but produces no detectable light. It did not, by itself, tell anyone what that mass actually is at the particle level. That question, whether dark matter consists of an entirely new type of subatomic particle, or something more exotic, or perhaps ordinary matter in a form simply too dim to detect (a possibility called MACHOs, discussed in the next chapter, which turned out to be largely ruled out by subsequent observations), remained wide open in the 1970s and, in important respects, remains open today. What Rubin accomplished was to transform dark matter from a marginal curiosity, associated with one difficult astronomer's disputed 1933 calculation, into a mainstream problem that the entire field of astrophysics recognized needed solving.

Rubin continued producing influential research for decades afterward, eventually becoming, in 1981, only the second woman ever elected to the National Academy of Sciences' astronomy section. She was a lifelong advocate for women in science, mentoring generations of younger astronomers and pushing, often against real institutional resistance, for greater inclusion of women in a field that had kept her own career on hold for the better part of a decade. She died in 2016, one year before the Nobel Prize committee would controversially decline to recognize her contribution to dark matter research with a prize of her own, a decision that remains, among many astronomers, a point of genuine and lasting frustration, since the Nobel Prize in Physics is not awarded posthumously and the window to honor her properly had, by then, permanently closed.

Her name lives on in a fitting way. In 2019, the large new astronomical survey telescope under construction in Chile, designed to conduct the most comprehensive optical survey of the sky ever attempted, was officially named the Vera C. Rubin Observatory. The telescope began science operations in the 2020s and is expected, over its planned decade-long survey, to generate some of the most precise maps yet of how dark matter is distributed across enormous volumes of the observable universe, extending the very line of inquiry Rubin herself opened with a spectrograph and a handful of photographic plates in the Arizona desert sixty years ago. There is a particular satisfaction in that: the instrument built to finally pin down the nature of dark matter carries the name of the astronomer whose stubborn attention to a strange printout first forced the world to take the problem seriously.

It is worth adding one further layer to Rubin's story before moving on, because it illustrates something about how scientific credit and scientific priority can diverge in ways that are worth understanding honestly rather than smoothing over. Rubin herself was consistently, publicly generous about acknowledging the earlier work her own findings built upon, regularly crediting both Zwicky's original 1933 insight and the radio astronomy measurements Morton Roberts had produced independently around the same time as her own early Andromeda results. This kind of explicit, public crediting of prior work matters more than it might seem, because the history of science is littered with cases where earlier contributions were quietly absorbed into a later, more famous name without proper acknowledgment, a pattern that can distort both the historical record and the incentives facing scientists deciding which problems are worth the risk of pursuing. Rubin's career, which began under real institutional discrimination and ended with her among the most honored astronomers of her generation, offers a useful case study in how a scientific field's memory of who deserves credit for a discovery can shift considerably over time, often lagging years or decades behind the moment the actual insight occurred, and often correcting only when someone within the field makes a sustained, public effort to set the historical record straight.

The next chapter turns to a question Rubin's work left wide open: if dark matter is real, and by the 1970s and 1980s few serious researchers doubted that it was, then what, physically, is it made of? The candidates that have been proposed, investigated, and in several important cases ruled out, form one of the more genuinely suspenseful ongoing investigations in modern physics, complete with false leads, clever experimental designs, and a shortlist of remaining suspects that has narrowed considerably, though not yet all the way down to one.

CHAPTER FOUR
The Suspects Lineup

Imagine a detective who has established, beyond reasonable doubt, that a crime occurred. The evidence is overwhelming: witnesses, physical traces, a consistent pattern across multiple independent investigations. What the detective does not yet have is a name. This is roughly where dark matter research stood by the late 1970s. Vera Rubin's rotation curves, combined with Zwicky's earlier cluster measurements and a growing body of supporting evidence, had established beyond serious scientific doubt that something with mass, lots of it, was out there, shaping the motion of galaxies. What nobody could say with confidence was what that something actually was, at the level of fundamental physics.

This chapter is about the suspects. Over the past five decades, physicists have proposed, tested, and in several cases eliminated a series of candidates for what dark matter might actually consist of. Some candidates were straightforward extensions of things we already knew existed. Others required inventing entirely new categories of particle, unlike anything in the Standard Model of particle physics, the well-tested framework that describes every particle and force physicists had confirmed as of the mid-1970s. Walking through this lineup, in roughly the order the field investigated and then largely set aside or continued pursuing each one, tells you a great deal about how particle physics and astrophysics collaborate, and about how a field narrows down an enormous space of theoretical possibilities using real observational constraints.

The first and most conservative suspect was, in a sense, the least imaginative: what if dark matter was not exotic at all, but simply ordinary matter in a form too dim to see? This category of candidate got the acronym MACHO, for Massive Compact Halo Object, and it covered things like brown dwarfs (failed stars too small to sustain nuclear fusion), black holes, neutron stars, and other dense but faint objects that could plausibly populate a galaxy's halo without producing enough light to show up in a normal survey. This was an attractive hypothesis because it required no new physics whatsoever. Every proposed MACHO candidate was already known to exist; the only open question was whether there were enough of them, scattered through galactic halos, to account for the missing mass Rubin and Zwicky had found.

Testing this hypothesis required a clever technique called gravitational microlensing, an idea proposed by the Polish astronomer Bohdan Paczynski in 1986. The logic works like this: if a massive but invisible object, a MACHO, happens to pass directly between Earth and a distant background star, its gravity will bend the light from that star, briefly focusing and brightening it in a predictable, temporary pattern, then fading back to normal as the MACHO moves out of alignment. This is a direct consequence of general relativity, the same theory that predicts gravitational waves, and it does not require the MACHO to emit any light of its own. You are using gravity itself as the detection mechanism.

Starting in the early 1990s, several international collaborations, including one called simply the MACHO Project, along with the French-led EROS collaboration and the Polish-led OGLE survey, began monitoring millions of stars in the Large Magellanic Cloud, a satellite galaxy of the Milky Way, watching for these brief gravitational lensing brightening events. If the Milky Way's dark matter halo were made substantially of MACHOs in the stellar mass range, these surveys should have detected a characteristic rate of lensing events. The MACHO Project did detect some events, and their initial analysis suggested MACHOs of roughly solar mass might account for perhaps twenty percent of the galactic halo. But the EROS and OGLE surveys, using larger stellar samples and more conservative statistical methods, found far fewer events than the MACHO Project's early results implied, and subsequent, more rigorous re-analyses narrowed the allowed MACHO contribution dramatically. By the mid-2000s, the combined weight of these microlensing surveys had essentially ruled out MACHOs across most plausible mass ranges as the primary explanation for dark matter. There remained, and still remains, a narrow allowed window for compact objects in a specific mass range (roughly ten to thirty times the mass of the Sun) to contribute some fraction of the halo, a window that has become newly interesting in the era of gravitational wave astronomy, since some of the black hole mergers detected by LIGO fall within exactly this mass range, raising the intriguing but still unconfirmed possibility that primordial black holes, formed in the extremely early universe rather than from the collapse of dying stars, might account for a portion of dark matter. This remains an active area of research rather than a settled question, and later in this book, when we discuss LIGO's black hole detections in detail, we will return to why some of those mergers surprised physicists in the first place.

For the most part, though, MACHOs as a complete explanation for dark matter did not survive contact with the data. This left the far larger and stranger category of candidates: entirely new types of fundamental particle, unlike anything in the Standard Model, existing specifically to solve the dark matter problem, or in some cases, existing because some other, unrelated theoretical problem in particle physics predicted them, with dark matter turning out to be a welcome bonus.

The leading candidate for decades has been something called a WIMP, short for Weakly Interacting Massive Particle. The name describes the proposed properties directly. A WIMP would have mass (unlike a photon, which is massless), and it would interact with ordinary matter only through the weak nuclear force and gravity, essentially never through electromagnetism, which is precisely why it would produce no light and prove so difficult to detect. WIMPs earned their status as the leading candidate largely because of something physicists call, with a bit of self-aware humor, the WIMP miracle. If you calculate how many particles with roughly the mass and interaction strength theorists were already proposing for entirely separate reasons (particles predicted by a class of theories called supersymmetry, developed originally to solve problems unrelated to dark matter) would have been produced in the hot, dense early universe, and then calculate how that population would thin out as the universe expanded and cooled, the resulting present-day abundance comes out remarkably close to the actual measured abundance of dark matter. This coincidence, arrived at from two entirely independent directions, cosmological observation on one side and particle theory on the other, converging on compatible numbers, is what earned WIMPs their favored status for so long.

The trouble is that decades of increasingly sensitive experiments, built specifically to detect WIMPs interacting with ordinary matter, have come back empty. This is not a small or casual research effort. It represents one of the more expensive and technically demanding branches of experimental physics operating today, and the next chapter of this book is devoted entirely to how these detectors work and what they have found (and, just as importantly, what they have ruled out). For now, the short version is this: as detector sensitivity has increased by many orders of magnitude since the 1990s, the parameter space in which a WIMP could plausibly be hiding, the range of possible masses and interaction strengths still consistent with observation, has shrunk dramatically. WIMPs have not been ruled out entirely. But the simplest, most theoretically favored versions of the WIMP hypothesis have been substantially constrained, pushing physicists to consider both more exotic variants of the WIMP idea and other candidates entirely.

The second major particle candidate is the axion, and its origin story is genuinely unusual, because the axion was not invented to explain dark matter at all. It was proposed in 1977 by the physicists Roberto Peccei and Helen Quinn to solve an entirely unrelated puzzle in particle physics called the strong CP problem, a technical issue involving why the strong nuclear force does not appear to violate a particular symmetry (called charge-parity, or CP, symmetry) even though, according to the equations governing quantum chromodynamics, there is no obvious reason it should not. The Peccei-Quinn solution to this problem predicted the existence of a new, extremely light particle, later named the axion, after a laundry detergent brand, in a bit of physicist humor, because the particle was proposed to clean up the strong CP problem. It was only afterward that physicists realized axions, if they exist, would also naturally behave as an excellent dark matter candidate: extremely light, extremely weakly interacting, and capable of being produced in exactly the right abundance in the early universe to account for the missing mass.

Axions are, in some ways, harder to search for than WIMPs, precisely because they are predicted to be so light and so weakly interacting that many of the direct detection techniques used for WIMPs simply do not apply. Instead, axion searches typically rely on a predicted, extremely subtle interaction between axions and magnetic fields, in which an axion passing through an intense magnetic field has a tiny probability of converting into a detectable photon. Experiments like the Axion Dark Matter eXperiment, run at the University of Washington, use exactly this principle, housing a resonant microwave cavity inside an extremely strong superconducting magnet, tuned carefully across a range of frequencies to search for the faint signal an axion conversion would produce. As of this writing, no confirmed axion detection has been made, but the experimental sensitivity continues to improve, and significant portions of the theoretically favored axion parameter space remain to be explored.

A third candidate, the sterile neutrino, comes from a different direction still. Ordinary neutrinos are already known, confirmed particles, produced constantly by nuclear reactions in the Sun and by countless other processes, and they interact so weakly with ordinary matter that trillions pass through your body every second without a single interaction. Sterile neutrinos would be a heavier, hypothetical cousin of these ordinary neutrinos, interacting even more weakly still, essentially only through gravity. Some extensions of the Standard Model, developed to explain why ordinary neutrinos have the small but nonzero mass experiments have confirmed they possess, naturally predict the existence of sterile neutrinos as a byproduct. If sterile neutrinos exist with the right mass range, typically discussed in the kiloelectronvolt range, they could account for some or all of dark matter, and they would be classified as warm dark matter, meaning particles that were moving at meaningfully high, though not relativistic, speeds when they decoupled from the rest of the universe's matter in its early history, a property that would leave subtly different fingerprints on the large scale structure of the universe compared to the colder, slower moving WIMPs. Searches for sterile neutrino dark matter typically look for a very specific type of decay signature, a particular X-ray emission line, in X-ray telescope observations of galaxy clusters and dwarf galaxies, and while some tentative signals have been reported over the years, none has held up to independent scrutiny as conclusive evidence.

There are other candidates further down the list, each with its own dedicated group of researchers and its own experimental search program: primordial black holes, formed not from dying stars but from extreme density fluctuations in the earliest fractions of a second after the Big Bang; a hypothetical heavy WIMP variant sometimes nicknamed the WIMPzilla; particles predicted by theories involving extra spatial dimensions beyond the three we experience directly. This book will not walk through every proposed candidate in exhaustive detail, both because the list continues to grow and because most readers will get more genuine understanding from a careful look at the leading candidates than from a superficial survey of every theoretical possibility ever proposed.

What is worth emphasizing, as this chapter closes, is the discipline the field has shown in narrowing this list through actual experimental constraint rather than theoretical fashion alone. Being an elegant idea, mathematically speaking, is not sufficient for a dark matter candidate to remain taken seriously. It has to survive contact with increasingly precise measurement, and both MACHOs and the simplest WIMP models have taken serious hits from exactly this kind of empirical pressure over the past three decades. This is science working as intended, even though it means the central question of this part of the book, what is dark matter actually made of, remains genuinely, honestly unanswered as of this writing.

It is worth pausing to consider why this uncertainty persists, given how much observational evidence supports dark matter's existence in the first place. The answer lies in a basic asymmetry between gravitational and non-gravitational evidence. Every line of evidence discussed so far in this book, Zwicky's cluster measurements, Rubin's rotation curves, and the gravitational lensing observations discussed in the next chapter, relies on gravity, and gravity, as far as any experiment has ever shown, does not care what kind of mass is producing it. A kilogram of ordinary matter and a kilogram of dark matter, whatever it turns out to be, produce identical gravitational effects. This is precisely why gravitational evidence can establish, with tremendous confidence, that dark matter exists and roughly how it is distributed, while remaining fundamentally silent on the question of composition. Determining what dark matter actually is requires catching it interacting through some channel other than gravity, however faint that interaction might be, which is exactly the painstaking work described in the next chapter.

There is also a sociological dimension to this hunt worth noting, because it shapes how seriously any given candidate gets pursued at any given time. Large experimental collaborations, the kind capable of building and operating a detector like the ones discussed in chapter five, require years of planning, tens or hundreds of millions of dollars in funding, and the sustained commitment of hundreds of scientists and engineers. Decisions about which candidates to prioritize, WIMPs, axions, sterile neutrinos, or some combination, are never made in a vacuum. They reflect a mixture of theoretical elegance, technological feasibility, and the accumulated, sometimes contentious, judgment of a research community about where the best odds of success actually lie. The WIMP hypothesis dominated funding and attention for so long partly because it was, for a genuine stretch of decades, the most theoretically compelling option available, supported by that striking coincidence between particle physics predictions and cosmological observation described earlier in this chapter. As WIMP searches have come back empty year after year, funding and talent have increasingly diversified toward axion searches and other alternatives, a shift you can track directly by looking at how experimental proposals and government physics funding priorities have evolved over the past fifteen years. Science, in other words, does not just test hypotheses. It also has to decide, collectively and imperfectly, which hypotheses are worth the enormous cost of testing rigorously, and that decision-making process is every bit as human, and occasionally as messy, as the discoveries that eventually result from it.

CHAPTER FIVE

Underground Hunters

Nearly a mile beneath the rolling grassland of Lead, South Dakota, in a former gold mine that once produced more gold than any other site in the Western Hemisphere, sits one of the most sensitive scientific instruments ever constructed. It does not look like much from a photograph: a cylindrical tank, roughly ten feet tall, filled with ultra-pure liquid xenon cooled to nearly one hundred degrees Celsius below zero, surrounded by layers of shielding, sensors, and support structure, the whole assembly submerged in a larger tank of purified water. This is the LUX-ZEPLIN experiment, known as LZ, and its purpose is almost absurdly patient: to sit quietly, for years at a time, waiting for a single dark matter particle to strike a single xenon atom, producing a flash of light and a tiny cascade of freed electrons too faint for almost any other instrument on Earth to notice.

The choice of location is not incidental. The old Homestake gold mine, now repurposed as the Sanford Underground Research Facility, provides something no laboratory on the surface can: nearly a mile of solid rock overhead, shielding the delicate detector from cosmic rays, the constant rain of high-energy particles from space that bombards the Earth's surface continuously. A cosmic ray striking a detector produces exactly the kind of signal you would also expect from a dark matter particle interaction, and at the surface, cosmic rays are common enough to overwhelm any hope of detecting the vastly rarer, vastly fainter signal a dark matter particle might produce. Bury the detector deep enough, and the rock overhead absorbs the overwhelming majority of cosmic radiation before it ever reaches the sensitive equipment below, dropping the background noise by many orders of magnitude and, for the first time, making it plausible to actually notice a signal as faint as a WIMP interaction would produce.

The physics behind how LZ actually works is worth walking through carefully, because the same basic principle underlies most of the world's leading dark matter detection experiments, including its close competitor, XENONnT, located under a mountain in central Italy at the Gran Sasso National Laboratory, and the Chinese-led PandaX-4T experiment. All three use the same core detector technology, a design called a dual-phase xenon time projection chamber, and understanding how it works tells you almost everything you need to know about the current state of direct dark matter detection.

Xenon, a heavy noble gas, was chosen as the detection medium for good physical reasons. Its high atomic mass makes it a relatively large, dense target for a hypothetical WIMP to collide with, increasing the odds of an interaction actually occurring, at least compared to using a lighter element. Xenon is also chemically inert (a noble gas, meaning it does not react easily with other substances) which makes it far easier to purify to the extraordinary standards these experiments require, since even trace contamination from naturally radioactive elements, elements present in nearly all ordinary materials in small quantities, would swamp the detector with false signals. And crucially, when a particle interacting with a xenon atom's nucleus causes that atom to recoil, the collision produces two nearly simultaneous, detectable signals: a flash of light (called scintillation) and a small number of freed electrons, both of which can be captured and measured with extraordinary precision.

Here is where the dual-phase design earns its name and its cleverness. The main body of the detector is filled with liquid xenon, but a thin layer at the very top is kept in gaseous form. When a particle interacts within the liquid xenon, it produces that first flash of light, detected almost instantly by arrays of extremely sensitive photomultiplier tubes positioned above and below the liquid. At the same time, the collision frees a small cloud of electrons, which drift slowly upward through the liquid xenon under the influence of a carefully maintained electric field, eventually crossing into the thin gas layer at the top, where they produce a second, distinct flash of light. By measuring the precise time delay between these two flashes, the initial scintillation flash and the delayed electron-drift flash, the detector can calculate exactly how deep within the liquid xenon the original interaction occurred, effectively giving the instrument three-dimensional positional information about every event it records, accurate to within a few millimeters.

This positional precision turns out to be one of the most powerful tools these experiments have for distinguishing a genuine dark matter signal from background noise. Real WIMPs, if they exist, are expected to interact so rarely, and so weakly, that any single WIMP would almost certainly strike only one xenon atom during its brief passage through the detector, producing a single, isolated interaction point. Background radiation, by contrast, whether from residual radioactive contamination in the detector materials or from stray cosmic ray particles that manage to penetrate the shielding, frequently produces multiple interaction points scattered through the detector volume, since these background particles tend to interact more readily with matter and often scatter multiple times before losing all their energy. By tracking the precise three-dimensional location of every recorded event, researchers can filter out multi-site interactions as background and focus their analysis on the rarer single-site events, which are far more consistent with what a genuine dark matter interaction would look like.

Even after this filtering, the vast majority of remaining single-site events are still background, not dark matter, which is why these experiments also measure the ratio between the two signals, the scintillation light and the electron-drift signal, since this ratio differs measurably depending on whether the original recoil was a nuclear recoil (a particle bumping directly into a xenon nucleus, which is what a WIMP interaction should look like) or an electron recoil (a particle interacting with the electron cloud surrounding a xenon atom, which is the far more common signature of ordinary background radiation). Combining three-dimensional positioning with this ratio discrimination allows researchers to reject the overwhelming majority of background events and focus their search on an extremely narrow, extremely rare category of signal: single-site nuclear recoils, occurring at a rate that, if WIMPs exist at all, might be expected to happen only a handful of times per year, in a detector containing many tons of xenon, watched continuously.

The scale of ultra-pure material required for these experiments is worth appreciating on its own terms. LZ uses more than eleven tons of liquid xenon, an extraordinary quantity given that xenon is among the rarer stable elements in Earth's atmosphere, present at a concentration of less than one part in ten million by volume. Producing eleven tons of xenon pure enough for this kind of experiment, virtually free of naturally radioactive contaminant isotopes like krypton-85 and radon, represents its own significant technical achievement, requiring specialized purification and distillation processes developed specifically for these detectors. The entire apparatus, water tank, xenon chamber, photomultiplier arrays, shielding, and support infrastructure, was assembled deep underground, transported piece by piece down a mile-deep mine shaft in a former gold mining elevator, and constructed in a cleanroom environment designed to keep out even microscopic dust particles that might introduce trace radioactive contamination.

So what have these enormously expensive, meticulously engineered instruments actually found? As of the most recent published results, the honest answer is: no confirmed dark matter detection, but a set of results that have transformed the field nonetheless. In results announced in December 2025, the LZ collaboration, an international team of more than 250 scientists and engineers from 37 institutions, reported on an analysis using 417 live days of data collected between March 2023 and April 2025, the largest dataset ever assembled by a dark matter direct detection experiment. Searching specifically for WIMPs in the low mass range between roughly three and nine times the mass of a proton, a region only recently made accessible by improvements in detector sensitivity, LZ found no evidence of dark matter interactions, and used the absence of a signal to place the tightest constraints yet on how strongly a WIMP in this mass range could possibly interact with ordinary matter, if it exists at all. The comparable XENONnT collaboration, using a slightly smaller xenon target but similarly refined analysis techniques, published its own leading constraints using a three point one tonne-year exposure in results reported in early 2025.

There is a genuinely fascinating and somewhat unexpected bonus that came out of the same December 2025 LZ results, one that illustrates something important about how sensitive these instruments have become. Buried within the same dataset, LZ researchers identified, for the first time in this particular experiment, a faint signal from neutrinos produced deep inside the Sun's core, specifically a category called boron-8 solar neutrinos, detected through a process called coherent elastic neutrino-nucleus scattering, a phenomenon that had only been directly observed for the first time in any experiment in 2017. The significance of this solar neutrino signal reached four point five standard deviations, comfortably past the three-sigma threshold physicists generally consider sufficient to claim tentative evidence, though still short of the five-sigma threshold conventionally required to claim a full discovery. This might sound like an unrelated curiosity, a detector built to find dark matter stumbling onto solar physics instead, but it actually points toward one of the central emerging challenges facing the entire field of direct dark matter detection: an effect researchers have started calling the neutrino fog.

The neutrino fog refers to a genuine and increasingly serious problem. As dark matter detectors like LZ and XENONnT become sensitive enough to probe lower and lower WIMP masses and weaker and weaker interaction strengths, they inevitably become sensitive enough to also detect a steady background of solar neutrinos, which produce a signal that, at these low energies, becomes difficult to distinguish from the very WIMP signal researchers are hunting for. Unlike ordinary radioactive background contamination, which can in principle be reduced through better purification and shielding, solar neutrinos cannot be shielded against. The Sun produces them constantly, they pass through the entire Earth as though it were not there, and no amount of rock overhead or material purification will ever screen them out. As detector sensitivity continues to improve in the coming years, researchers increasingly expect to run directly into this neutrino fog, a fundamental physical background that may ultimately limit how sensitive xenon-based WIMP searches can become using current detector designs.

This has already begun shaping how the field plans its next generation of instruments. Many of the same researchers currently working on LZ are involved in designing a successor experiment called XLZD, intended to combine the best engineering lessons from LZ, XENONnT, and a related planned experiment called DARWIN into a single next-generation liquid xenon observatory, one large enough and sensitive enough to push WIMP searches to their practical limit while also opening new scientific opportunities in neutrino physics and other rare-event searches, essentially treating the neutrino fog not purely as an obstacle but as a scientific resource in its own right.

It would be reasonable, reading this chapter, to wonder whether decades of null results across increasingly expensive and sophisticated experiments should be read as evidence against the WIMP hypothesis altogether, rather than merely as evidence that the search needs refining. This is a genuinely live question within the physics community, and reasonable, well-informed researchers disagree about how to weigh it. What is not in serious dispute is this: these experiments, even without a confirmed detection, have already accomplished something scientifically valuable. Each new null result rules out a specific, previously plausible region of WIMP mass and interaction strength, narrowing the space of remaining possibilities with each passing year, the same way a detective's investigation narrows a suspect list even when each individual lead turns out to be a dead end. Somewhere in that shrinking remaining space, or perhaps in an entirely different category of candidate discussed in the previous chapter, the actual identity of dark matter is still waiting to be found. The underground hunters of South Dakota, Italy, and China have not yet caught their suspect. But with every passing year of data collection, the net they are using grows finer, and the places dark matter could still be hiding grow fewer.

It is worth taking a moment to appreciate the sheer engineering discipline required to keep an instrument like LZ running reliably for years at a time, a mile underground, in an environment about as far removed from a typical physics laboratory as one could design on purpose. Every component that goes into the detector, every bolt, cable, and seal, has to be individually screened for radioactive contamination before installation, because even trace amounts of naturally occurring uranium or thorium in ordinary construction materials can produce enough background radiation to swamp the signal researchers are trying to isolate. The water tank surrounding the central xenon chamber serves a similar shielding purpose, absorbing residual radiation from the surrounding rock itself, since even a mile of granite is not perfectly clean of natural radioactivity. Temperature within the xenon chamber has to be maintained with remarkable stability, because even small thermal fluctuations can alter the xenon's density and therefore the precision of the timing measurements the detector depends on. None of this is glamorous work, in the sense that it rarely produces a dramatic headline on its own. But it is precisely this unglamorous, meticulous engineering that makes the difference between an instrument capable of detecting a signal a thousand times fainter than any previous generation of detector and one that simply drowns in its own background noise. The history of dark matter detection, at the level where the actual work happens, is substantially a history of getting quieter: building instruments with less inherent noise, purifying materials to greater standards, and shielding against every conceivable source of interference, one careful engineering decision at a time.

CHAPTER SIX
A Universe Flying Apart

By 1997, two teams of astronomers were engaged in what amounted to a friendly, occasionally tense race, though almost nobody outside professional astronomy was paying attention. The prize was not fame, at least not initially. It was a number: the deceleration parameter, a single figure describing how quickly the expansion of the universe was slowing down under the pull of its own gravity. Both teams fully expected to find that the universe's expansion, set in motion by the Big Bang roughly thirteen point eight billion years ago, was gradually decelerating, the same way a ball thrown upward slows as gravity pulls it back down. The only real question, as far as either team anticipated, was by how much.

One team, the Supernova Cosmology Project, was led by Saul Perlmutter, an astrophysicist at Lawrence Berkeley National Laboratory who had been building toward this measurement since the late 1980s. The rival team, the High-z Supernova Search Team, was led by Brian Schmidt, formed in 1994 partly in response to Perlmutter's group, using broadly similar methods but with an independent team, independent telescope time, and, crucially, independent data. Working within Schmidt's team, a young researcher named Adam Riess led the detailed analysis that would ultimately produce the team's headline result.

Both groups were using the same basic technique, one that depends on a peculiar and extraordinarily useful class of stellar explosion called a Type Ia supernova. When a white dwarf star, the dense, burnt-out remnant of a Sun-like star, accumulates too much mass, typically by pulling material from a companion star in a binary system, it crosses a critical threshold called the Chandrasekhar limit and detonates in a runaway thermonuclear explosion, briefly outshining its entire host galaxy. What makes Type Ia supernovae so valuable to cosmologists is that this explosion mechanism is remarkably consistent from one occurrence to the next, since it is triggered by the same physical threshold every time, which means these explosions reach very similar peak brightness across different events, once a small correction is applied for the small variations that do occur, a correction Perlmutter's team and others had spent years refining. This consistency makes Type Ia supernovae what astronomers call standard candles: objects of known, predictable brightness that can be used to calculate distance, since an object's apparent brightness as seen from Earth depends predictably on both its true brightness and its distance, the same way a lighthouse looks dimmer the farther away you are, in a mathematically precise, calculable way.

The strategy, then, was to find as many distant Type Ia supernovae as possible, measure their apparent brightness and their redshift (how much their light has been stretched toward the red end of the spectrum by the expansion of the universe during its journey to Earth, which tells you how much the universe has expanded since the light was emitted), and plot the relationship between the two. If the universe's expansion were decelerating, as both teams expected, distant supernovae at a given redshift should appear brighter than a universe expanding at a constant rate would predict, because the deceleration means the universe was expanding faster in the past than a simple constant-rate model would suggest, putting those ancient supernovae effectively closer to us, in terms of the light-travel distance, than they would otherwise be.

Finding a Type Ia supernova, let alone one distant enough to be useful for this measurement, is genuinely difficult. These events are rare, unpredictable, and visible only for a matter of weeks before fading beyond detection. Both teams developed techniques for scanning large areas of sky repeatedly, comparing images taken weeks apart to spot new points of light that had not been there before, a laborious process that, in the pre-automated era of the 1990s, required enormous amounts of dedicated telescope time and painstaking manual verification. Once a candidate supernova was identified, teams needed to move quickly to obtain a spectrum confirming it was genuinely a Type Ia explosion and not some other, less useful type of stellar event, all before the object faded from visibility.

By 1997 and into 1998, both teams had accumulated enough distant supernovae to begin producing meaningful results, and this is where the story takes its now-famous turn. Rather than confirming a decelerating universe, the data from both teams, gathered independently, using overlapping but not identical samples of supernovae, analyzed with independently developed statistical methods, pointed to the same startling, unanticipated conclusion: distant supernovae were dimmer than a decelerating universe model predicted. Not brighter, as expected from deceleration. Dimmer. This meant those supernovae were farther away than a decelerating expansion model would place them, which meant the universe's expansion had not been slowing down. It had been speeding up.

It is worth dwelling on just how surprising this was, because the surprise is central to why the discovery carried such weight. Every researcher on both teams had begun this project expecting to measure some degree of deceleration. The mathematics of general relativity, applied to a universe filled with ordinary and dark matter, predicts that gravity should act as a brake on cosmic expansion over time, gradually slowing the rate at which galaxies recede from one another. Finding acceleration instead required either a serious flaw in the measurement, or a serious flaw in the underlying assumption that gravity, sourced purely by matter, was the only thing shaping the universe's large-scale expansion. Both teams spent considerable time in 1997 and 1998 checking and rechecking their analyses, hunting for the mistake they assumed must be there, because an accelerating universe implied something genuinely radical: some kind of repulsive effect, acting on cosmic scales, counteracting and overwhelming the attractive pull of gravity from all the matter, dark and ordinary combined, in the universe.

The High-z team's first results, based on an initial sample of ten distant supernovae, were published in the Astronomical Journal in 1998, with Riess as lead author. The Supernova Cosmology Project published its own larger analysis, based on forty two supernovae, in the Astrophysical Journal the following year, in 1999. Despite using different supernova samples, different telescopes, and independently developed analysis pipelines, both papers reached the same conclusion, within statistical uncertainty: the expansion of the universe is accelerating, not decelerating, and whatever is driving that acceleration appears to account for roughly seventy percent of the total mass-energy content of the universe.

This kind of independent convergence, two competing teams, motivated in part by rivalry, reaching the same startling conclusion through separately gathered data, is close to the gold standard of scientific evidence. It is much harder to dismiss a surprising result as an error, a fluke, or a shared methodological blind spot when two independent groups, who would have been thrilled to catch the other team in a mistake, instead confirm each other's central finding. The scientific community responded accordingly. What might have been dismissed as a curious anomaly from a single research group instead rapidly reshaped the central working model of cosmology.

The substance that appeared to be driving this accelerating expansion needed a name, and cosmologists settled on dark energy, deliberately echoing the earlier terminology of dark matter, though the two phenomena are, as far as anyone currently understands, entirely unrelated apart from both being invisible and both being inferred rather than directly observed. Dark matter, discussed across the preceding four chapters of this book, exerts ordinary gravitational attraction, the same as any other mass, and it clusters together with galaxies and galaxy clusters, shaping their structure and rotation. Dark energy behaves in the opposite manner: rather than pulling things together, it appears to push space itself apart, and unlike dark matter, it appears to be distributed smoothly and uniformly throughout the universe, not clustered around galaxies at all, a property that turns out to be an important clue about its underlying nature, one we will pick up in the next chapter.

The Nobel Prize in Physics for 2011 recognized this discovery, awarded jointly to Saul Perlmutter, Brian Schmidt, and Adam Riess for what the Nobel committee's citation described as the discovery of the accelerating expansion of the universe through observations of distant supernovae. It is worth noting, as this book will note again in later chapters, how relatively quickly this particular discovery moved from initial announcement to Nobel recognition, thirteen years, a comparatively short interval by the standards of physics Nobels, reflecting both the scientific significance of the finding and the strength of the evidence supporting it, gathered independently by two competing teams whose rivalry, in the end, served the deeper interests of getting the science right.

It is worth pausing on the human dimension of this story before moving forward, because the rivalry between the two teams was genuine, not a polite fiction constructed after the fact for the benefit of popular science writing. Perlmutter's Supernova Cosmology Project had a multi year head start, having begun developing its supernova search techniques in the late 1980s, and Schmidt's High-z team was assembled in 1994 in part because researchers who felt Perlmutter's group was moving too slowly, or was insufficiently open to outside collaboration, wanted an independent effort capable of reaching a result on its own timeline. The competition between the two groups was real enough that both teams raced to process and interpret their data quickly once the surprising acceleration signal began to emerge, each aware that the other might reach a public announcement first. This kind of competitive pressure is sometimes viewed skeptically, as a corrupting influence on careful science, and there are certainly documented cases across the history of research where competitive haste has led to premature or sloppy claims. But in this particular instance, the competition appears to have functioned as an additional safeguard rather than a liability, precisely because neither team had any incentive to simply validate the other's mistake. If Perlmutter's team had made a subtle error, Schmidt's team, eager to demonstrate the value of its own independent approach, had every reason to catch it, and vice versa. That neither team found a disqualifying flaw in the other's work, despite considerable professional incentive to look for one, adds meaningfully to the overall credibility of the finding.

There is a broader lesson embedded in this chapter's story about how major discoveries actually tend to happen in physics, one that will recur throughout the remainder of this book. Neither team set out to discover dark energy. Both teams set out to measure something else entirely, the rate at which cosmic expansion was decelerating, confident in what they expected to find. The discovery emerged not from a hypothesis specifically designed to detect dark energy, but from careful, rigorous measurement of something else, followed by an honest, if initially reluctant, acceptance of what the data actually showed rather than what theory had predicted it should show. This is, in many ways, science working at its best: measure carefully, check your work relentlessly when the result surprises you, and when the surprising result survives every check you can throw at it, have the intellectual honesty to follow the evidence rather than your prior expectations, however uncomfortable that may be.

The discovery of cosmic acceleration also did something else, something worth pausing on before moving to the next chapter. It completed a picture that had been forming, piece by piece, since Zwicky's 1933 calculation. Ordinary visible matter, the stuff of stars, planets, and this very book, accounts for roughly five percent of the universe's total mass and energy content. Dark matter, established through the rotation curves and cluster dynamics discussed in earlier chapters, accounts for roughly twenty seven percent. And dark energy, the subject of this chapter and the next, accounts for the remaining sixty eight percent, the single largest component of the universe by a wide margin, and also, in an important sense, the most mysterious.

It is worth explaining, in a bit more technical detail than the discovery narrative above required, exactly how a plot of supernova brightness against redshift actually reveals something as abstract as cosmic acceleration, because the logic, once you see it clearly, is one of the more satisfying pieces of reasoning in modern astrophysics. Redshift, denoted by astronomers with the letter z, measures how much the wavelength of light has been stretched during its journey from a distant object to Earth, a stretching caused directly by the expansion of space itself during the time the light was in transit. A supernova with a redshift of z equals 0.5, for example, has light that has been stretched so that its wavelengths are fifty percent longer than when they were emitted, which tells you, given a model for how the universe has expanded over time, roughly how long ago that light left its source and roughly how much the universe has grown in the meantime.

Apparent brightness, separately, depends on the actual distance light has had to travel, specifically what cosmologists call luminosity distance, a quantity that is sensitive to the entire expansion history of the universe between the time the light was emitted and the time it arrives at our telescopes. This is the crucial subtlety: two different cosmological models, one with a decelerating expansion history and one with an accelerating expansion history, will predict different luminosity distances for a supernova at the same redshift, even though redshift alone does not distinguish between the two models. By comparing observed apparent brightness (which reveals actual luminosity distance) against redshift (which reveals how much the universe has expanded since the light was emitted) across a large sample of supernovae spanning a range of distances, Perlmutter's and Schmidt's teams were able to reconstruct, essentially, a history of how the universe's expansion rate has changed over billions of years. And that reconstructed history showed expansion speeding up, not slowing down, precisely in the more recent cosmic past.

There is a useful, if imperfect, everyday analogy for what the two teams found. Imagine tracking a car's speed using nothing but photographs taken along its route, with no direct speedometer reading available. If you know the car left home at a certain time and measure how far it has traveled by certain points, you can infer whether it was accelerating, decelerating, or moving at constant speed along the way, purely from the pattern of distances and times, without ever directly observing velocity itself. Perlmutter's and Schmidt's teams did something conceptually similar for the universe as a whole, using supernova brightness as a proxy for distance and redshift as a proxy for time elapsed since emission, reconstructing an expansion history that neither team expected and that neither team could initially believe.

It is also worth naming, honestly, some of the sources of uncertainty and controversy that surrounded this discovery in its early years, because good science writing does not pretend a landmark result arrived free of legitimate scrutiny. Critics in the late 1990s and early 2000s raised reasonable concerns about whether distant Type Ia supernovae might differ systematically from nearby ones in ways that could bias the brightness measurements, since a supernova exploding billions of years ago formed from stellar material with a different chemical composition than more recent supernovae, potentially altering the explosion's peak brightness in subtle ways. Researchers also worried about intervening dust, scattered along the vast distances between distant supernovae and Earth, which could dim supernova light in ways that might mimic the signature of accelerating expansion. Over the following decade, extensive follow-up work, including larger supernova samples, improved calibration techniques, and entirely independent lines of evidence discussed in the next chapter, addressed these concerns one by one, and the finding of cosmic acceleration has only grown more solidly established with each additional line of confirming evidence, rather than weakening under continued scrutiny, which is precisely the pattern you want to see from a genuine discovery rather than a statistical artifact.

We will spend the next chapter exploring exactly what dark energy might actually be, an investigation that will take us from Einstein's own equations, written decades before the discovery of cosmic acceleration, to the current, honestly unresolved debate about whether dark energy is a fixed, unchanging property of empty space itself, or something stranger still, evolving subtly over cosmic time in ways current observations are only just beginning to probe.

CHAPTER SEVEN

Meet Dark Energy

There is a strange historical irony sitting at the center of dark energy's story, and it is worth starting there rather than saving it for later. Decades before anyone had observed cosmic acceleration, Albert Einstein had already written down the mathematical tool that would eventually be used to describe it, and he had already decided he regretted doing so.

In 1917, roughly a year after publishing his general theory of relativity, Einstein tried applying his new equations to the universe as a whole, attempting to construct a mathematical model of the cosmos. He ran into a problem. His equations, applied to a universe filled with ordinary matter and nothing else, refused to produce a stable, static universe. Gravity, being purely attractive under the physics known at the time, would inevitably cause any such universe to either collapse inward or, if given an initial push, continue expanding or contracting indefinitely. Since the prevailing assumption in 1917, shared by essentially every astronomer and physicist of the era, was that the universe was static, neither expanding nor contracting, eternal and unchanging on the largest scales, Einstein needed his equations to produce that result. So he added a new term to his field equations, a constant value he called the cosmological constant, denoted by the Greek letter lambda, engineered specifically to introduce a repulsive effect that could exactly balance the attractive pull of gravity and hold a static universe in place.

Twelve years later, in 1929, the American astronomer Edwin Hubble published observations demonstrating that distant galaxies are, in fact, receding from us, and the farther away a galaxy is, the faster it appears to be receding, a relationship now known as Hubble's Law. This was unambiguous evidence that the universe is not static at all. It is expanding. With this evidence in hand, Einstein's original motivation for introducing the cosmological constant evaporated, and according to a widely repeated account, sourced originally to the physicist George Gamow's recollections rather than to any of Einstein's own writings, Einstein came to regard the cosmological constant as the biggest blunder of his career. Whether Einstein used precisely that phrase remains a subject of some historical debate among physicists and historians of science, since the quote traces back to Gamow's secondhand report rather than a documented statement from Einstein himself, but the underlying sentiment, that introducing the constant had been an unnecessary and ultimately embarrassing detour, is well supported by Einstein's own later writings on the subject, in which he described the term as theoretically unjustified and unnecessary once a dynamic, expanding universe was accepted.

For nearly seventy years, the cosmological constant remained a discarded footnote in the history of physics, a cautionary tale about letting philosophical preference (in Einstein's case, a preference for a static, unchanging cosmos) distort the interpretation of one's own equations. Then, in 1998, Perlmutter's and Schmidt's teams discovered that the universe's expansion was not merely continuing but accelerating, and physicists found themselves needing exactly the kind of repulsive, space-stretching effect Einstein had once proposed and then abandoned. The cosmological constant, exiled from serious physics for most of a century, was suddenly relevant again, not because anyone had rehabilitated Einstein's original static-universe motivation, but because it turned out to be one of the simplest, most mathematically natural ways of describing the newly discovered acceleration within the existing framework of general relativity.

So what, physically, does the cosmological constant actually represent? This is where dark energy connects to one of the strangest ideas in modern physics: the notion that empty space itself, the vacuum, is not truly empty, but carries an intrinsic energy density, an amount of energy present in every unit of volume of space regardless of whether any matter or radiation occupies that space. This is not an intuitive idea, and it is worth taking slowly. In everyday experience, and in classical physics generally, energy is something particles and fields carry. Remove all particles and fields from a region of space, and you would expect the energy content of that region to drop to zero. Quantum field theory, the framework underlying the Standard Model of particle physics, predicts otherwise. According to quantum field theory, even a perfect vacuum, containing no particles whatsoever, still possesses a baseline energy, arising from quantum fluctuations that are always present, even in a state you would naively describe as completely empty. Chapter eight of this book explores this idea, called vacuum energy, in considerably more depth, including a laboratory-measurable effect named after the physicist who predicted it. For now, the key point is this: if empty space carries an intrinsic energy density, and if general relativity is correct in treating all forms of energy as sources of gravity, then a universe filled with vacuum energy should behave in a specific, calculable way, one that turns out to produce exactly the kind of accelerating expansion the supernova surveys detected.

Here is the crucial and genuinely counterintuitive physics that makes this work. Ordinary matter and radiation, as the universe expands and a given region of space grows larger, become diluted, spread across a greater volume, and their energy density (energy per unit volume) correspondingly decreases. This is intuitive: spread the same amount of stuff across more space, and the concentration goes down. Vacuum energy behaves entirely differently. Because it is a property of space itself, rather than a fixed quantity of stuff distributed within space, vacuum energy density remains constant even as space expands. As the universe grows larger, there is simply more space, and therefore more total vacuum energy, even though the energy density within any given cubic meter of that space stays exactly the same. Combined with a further consequence of general relativity, namely that a form of energy with sufically negative pressure exerts a repulsive rather than attractive gravitational effect, a constant vacuum energy density, spread uniformly throughout ever-expanding space, produces precisely the self-reinforcing acceleration the supernova data revealed: expansion that not only continues but speeds up over time, because the accelerating effect grows in absolute terms as the volume of space itself grows.

This picture, dark energy as a manifestation of vacuum energy mathematically equivalent to Einstein's revived cosmological constant, is currently the leading and most observationally successful model, generally referred to by cosmologists as the Lambda-CDM model, where Lambda represents the cosmological constant and CDM stands for cold dark matter, the dark matter candidate framework discussed in chapters three through five. Lambda-CDM has proven remarkably successful at explaining an enormous range of independent cosmological observations using a comparatively small number of free parameters, and it currently represents the standard working model of cosmology, taught in graduate programs and used as the baseline framework against which new observations are typically compared.

But the story does not end there, and intellectual honesty requires acknowledging a genuine and unresolved tension sitting at the heart of this model. When physicists attempt to actually calculate the expected vacuum energy density from first principles, using the equations of quantum field theory, the predicted value comes out spectacularly, almost absurdly larger than the actual observed value of dark energy density inferred from cosmological measurements, by a factor that has been described, without much exaggeration, as the worst prediction in the history of physics, off by something on the order of sixty to one hundred and twenty orders of magnitude depending on precisely how the calculation is framed. This mismatch, known in the physics literature as the cosmological constant problem, remains one of the deepest unsolved puzzles in fundamental theoretical physics, a genuine embarrassment sitting beneath an otherwise remarkably successful cosmological model. Nobody currently has a fully satisfying explanation for why the actual, observed value of dark energy density is so many orders of magnitude smaller than naive quantum field theory calculations suggest it ought to be, though it is not for lack of serious theoretical effort; proposed resolutions range from subtle cancellations between different quantum contributions that current theory cannot yet properly calculate, to more radical ideas involving alternate universes or fundamentally different formulations of quantum gravity, a theory that does not yet exist in complete form.

There is also a live observational question worth flagging honestly here, one that active research programs are currently working to resolve: is dark energy actually constant over cosmic time, as the simplest cosmological constant model assumes, or does it evolve, weakening or strengthening as the universe ages? This distinction matters enormously for understanding the universe's ultimate fate, discussed briefly at the end of this chapter, and it has become an active area of genuine observational uncertainty in the past several years. Large-scale surveys, including the Dark Energy Spectroscopic Instrument survey conducted using a telescope in Arizona, have in recent years produced results hinting, at a still-preliminary level of statistical confidence, that dark energy's strength might not be perfectly constant after all, potentially varying subtly over the history of the universe rather than remaining fixed as Einstein's original cosmological constant would require. These hints remain provisional, requiring additional data and independent confirmation before the field will treat them as established, but they represent exactly the kind of genuinely open question that makes current cosmological research an active, unsettled field rather than a finished body of settled knowledge.

If dark energy does turn out to be a true, unchanging cosmological constant, the long-term implications for the universe's fate are sobering, in a quiet, almost geological sort of way. Continued, ever-accelerating expansion means that, over sufficiently long timescales, distant galaxies will eventually recede from us faster than light can bridge the growing gap, effectively vanishing from our observable universe. Cosmologists studying this scenario have calculated that within roughly a hundred billion years, an observer on Earth (or whatever remains of Earth by then) would be able to see only a few hundred nearby galaxies, gravitationally bound to our own Local Group, rather than the hundreds of billions currently visible with modern telescopes. This is sometimes called the heat death or the Big Freeze scenario: not a dramatic collapse, but a slow, inexorable dilution, as an accelerating universe spreads its remaining matter and energy across an ever-expanding volume of space, eventually leaving nothing but cold, dark, isolated remnants, adrift in a universe grown too vast and too empty to sustain the complex structures, stars, galaxies, and, presumably, any form of life, that briefly flourished within it.

That is, admittedly, a bleak note to end a chapter on, and it is worth remembering that this projected fate lies not merely millions but many tens of billions of years in the future, a timescale so vast it strains ordinary human intuition to even attempt comprehending it.

Before moving on, it is worth explaining a little more about how cosmologists actually pin down the specific numbers mentioned throughout this chapter, since the precision involved is genuinely impressive and worth appreciating. The current best estimate, that dark energy accounts for roughly sixty eight percent of the universe's total mass-energy content, with dark matter contributing roughly twenty seven percent and ordinary matter the remaining five percent, comes not from the supernova measurements alone but from combining multiple, entirely independent observational methods, each sensitive to different aspects of cosmic history. The supernova distance measurements described in the previous chapter provide one line of evidence. A second, entirely independent line comes from detailed studies of the cosmic microwave background, the faint afterglow of radiation left over from roughly 380,000 years after the Big Bang, when the universe first became transparent to light. Tiny temperature variations in this background radiation, mapped with extraordinary precision by satellite missions including NASA's WMAP mission and the European Space Agency's Planck mission, encode information about the universe's geometry, composition, and expansion history, and analyzing these variations produces independent estimates of the dark energy, dark matter, and ordinary matter fractions that agree remarkably well with the supernova-based measurements. A third line of evidence comes from studying the large-scale distribution of galaxies themselves, using a subtle statistical pattern called baryon acoustic oscillations, essentially fossilized sound waves from the early universe imprinted on how galaxies are distributed across cosmic distances, which provides yet another independent geometric probe of cosmic expansion history. When three genuinely independent methods, supernova distances, cosmic microwave background patterns, and large-scale galaxy clustering, converge on compatible values for the universe's composition, the resulting picture earns a level of confidence that no single method could provide alone, which is precisely why the Lambda-CDM model, despite the deep theoretical puzzle of the cosmological constant problem, remains the most observationally successful cosmological framework currently available.

The Dark Energy Spectroscopic Instrument, mentioned briefly above and generally referred to by its acronym DESI, deserves a bit more explanation, since it represents the current observational frontier in this particular investigation. Mounted on the Mayall four meter telescope at Kitt Peak National Observatory in Arizona, the same observatory where Vera Rubin conducted some of her pioneering rotation curve measurements decades earlier, DESI is designed to measure the positions and distances of tens of millions of galaxies and quasars, building the most detailed three-dimensional map of the universe's large-scale structure ever assembled. By tracking how the pattern of galaxy clustering has evolved across roughly eleven billion years of cosmic history, DESI provides a uniquely powerful tool for testing whether dark energy's strength has remained genuinely constant over that enormous span of time, or whether, as some of the collaboration's early results have tentatively suggested, it may have varied subtly, an unresolved question the DESI team continues to investigate with each successive year of additional data collection. Whatever the ultimate answer turns out to be, and it remains a genuinely open question as of this writing, resolving it will represent one of the most consequential results in cosmology in decades, either confirming Einstein's simplest mathematical picture of the vacuum or requiring physicists to develop an entirely new theoretical framework for understanding the dominant component of the universe.

The next chapter turns from this cosmic-scale mystery to something that might seem, at first glance, entirely unrelated: the strange, measurable energy of empty space itself, examined not at the scale of the observable universe but at the scale of a laboratory tabletop, where physicists have found ways to measure the restless quantum foam of the vacuum directly, using nothing more exotic than two carefully positioned metal plates.

CHAPTER EIGHT

The Empty Space That Isn't Empty

Take two ordinary metal plates. Make sure they carry no electric charge. Position them parallel to each other, separated by a gap smaller than the width of a human hair, and place the whole arrangement in a vacuum, a region with no air, no gas, nothing that could conduct force between them through ordinary physical contact. Classical physics, the physics of Newton and Maxwell, makes an unambiguous prediction about what happens next: nothing. With no charge and no material medium between them, the plates should simply sit there, entirely indifferent to each other's presence, apart from an utterly negligible pull of ordinary gravity.

Quantum physics makes a different prediction. It says the plates will be pulled together, drawn by a real, measurable force, arising from nothing more than the empty space between them.

This is the Casimir effect, predicted in 1948 by the Dutch physicist Hendrik Casimir while he was working, of all places, at the Philips Research Laboratories in Eindhoven, a commercial research facility better known at the time for developing consumer electronics than for foundational contributions to quantum field theory. Casimir arrived at his prediction while investigating an entirely different, more mundane problem, the behavior of colloidal solutions, mixtures where microscopic particles are suspended within a liquid, a topic with genuine industrial relevance to Philips's business interests. In the course of that applied research, working alongside his colleague Dirk Polder, Casimir found himself needing to understand how the quantum vacuum behaves near a conducting surface, and the calculation led him somewhere he had not expected: toward a testable prediction that two neutral, uncharged plates, placed close enough together in a vacuum, should experience a small but genuine attractive force, drawing them toward each other.

To understand why, you need to sit with an idea that sounds, on first encounter, like a contradiction in terms: the vacuum is not empty. According to quantum field theory, the framework physicists use to describe how particles and forces behave at the most fundamental level, even a region of space containing no particles whatsoever, no photons, no electrons, nothing that could be measured or detected directly, still seethes with activity. This activity takes the form of what physicists call virtual particles: pairs of particles and their corresponding antiparticles that spontaneously appear out of the vacuum, exist for an extraordinarily brief instant, and then annihilate each other, disappearing again, all within a timescale so short that they can never be directly observed or measured as ordinary particles. This is not a fringe or speculative idea. It follows directly from a foundational principle of quantum mechanics called the Heisenberg uncertainty principle, which places a fundamental limit on how precisely certain pairs of physical quantities, including energy and time, can simultaneously be known. One consequence of this uncertainty is that empty space can briefly violate what looks like strict energy conservation, borrowing a small amount of energy to create a fleeting virtual particle pair, provided that energy is returned, the particles reannihilated, quickly enough to satisfy the uncertainty principle's constraint. The vacuum, understood this way, is less like an empty stage and more like a restless, foaming sea, constantly generating and reabsorbing ripples too fast and too faint for ordinary detection.

Casimir's insight was to realize that this restless quantum foam could be manipulated, at least in a limited, indirect way, by introducing physical boundaries into the vacuum. Picture the space between two parallel conducting plates. In principle, the vacuum in that gap should support virtual photon fluctuations of every possible wavelength, the same as vacuum anywhere else. But a conducting plate imposes a physical constraint: an electromagnetic wave reflecting off a perfect conductor must have zero amplitude exactly at the conductor's surface, a boundary condition that rules out certain wavelengths entirely. Only virtual photon wavelengths that fit a whole number of times into the gap between the two plates are permitted to exist there, in the same way a guitar string of fixed length can only vibrate at specific resonant frequencies, not at any frequency whatsoever. Outside the plates, in the vast open vacuum with no boundary constraint, the full, unrestricted spectrum of virtual photon wavelengths remains available.

This creates an imbalance. The restricted region between the plates supports fewer possible virtual photon modes than the unrestricted region outside, meaning the vacuum energy density inside the gap is slightly lower than the vacuum energy density outside. And just as a pressure difference between two regions of gas produces a net force pushing from high pressure toward low pressure, this vacuum energy density imbalance produces a net force, and Casimir's calculation showed that force acts to push the plates together, drawing them from the higher-energy exterior region toward the lower-energy interior gap. The closer the plates are positioned, the fewer wavelengths fit in the gap, the stronger the imbalance becomes, and the stronger the attractive force grows, a relationship Casimir worked out with mathematical precision, predicting a force on the order of a few hundred piconewtons (an almost inconceivably small unit of force) when the plates are held roughly one hundred nanometers apart, a separation smaller than most bacteria.

For nearly fifty years after Casimir's 1948 prediction, actually measuring this force with any real precision remained beyond the reach of available laboratory technology. An early attempt in 1958, conducted by Casimir's Philips colleague Marcus Sparnaay using a spring balance and a capacitance measurement, produced results broadly consistent with Casimir's prediction but with error margins far too large to count as a rigorous confirmation. It was not until 1997 that the physicist Steve Lamoreaux, then working at the University of Washington and later at Los Alamos National Laboratory, conducted an experiment precise enough to genuinely test Casimir's fifty-year-old prediction, using a modern torsion pendulum apparatus capable of detecting forces far smaller than anything available to Sparnaay in the 1950s. Lamoreaux's measurements matched Casimir's theoretical prediction to within about five percent, a genuinely convincing confirmation, and subsequent experiments over the following decades have refined that agreement further still, with modern Casimir force measurements now achieving sub-percent precision, matching theory to a remarkable degree of accuracy.

The Casimir effect stands, as of this writing, as one of the most direct and unambiguous experimental confirmations available that quantum vacuum fluctuations are not merely a convenient mathematical bookkeeping device, but a genuine physical reality, capable of producing measurable, real-world forces between ordinary objects. This matters enormously for the broader story this book is telling, because it provides laboratory-scale, repeatable, tabletop evidence for the same basic physical principle underlying the dark energy discussion in the previous chapter: that empty space carries real energy, and that energy has real, measurable physical consequences.

It is worth being precise here about what the Casimir effect does and does not demonstrate in relation to dark energy, since popular explanations sometimes blur this distinction in ways that overstate the connection. The Casimir effect demonstrates, convincingly, that vacuum fluctuations are real and that boundary conditions can alter the local vacuum energy density in measurable, calculable ways. It does not, by itself, resolve the cosmological constant problem discussed in the previous chapter, the enormous, still-unexplained mismatch between the vacuum energy density predicted by quantum field theory calculations and the much smaller value of dark energy density actually observed on cosmic scales. In fact, in something of an ironic twist, the same basic vacuum fluctuation calculations that correctly predict the Casimir force between two nearby plates, when extended to predict the total vacuum energy density of the universe as a whole, produce a number roughly one hundred and twenty orders of magnitude larger than the actual measured dark energy density, the same catastrophic mismatch discussed in chapter seven. The Casimir effect, in other words, confirms that vacuum energy is real and behaves in ways quantum theory correctly predicts at laboratory scales, while simultaneously deepening rather than resolving the mystery of why the universe's actual dark energy density is so many orders of magnitude smaller than naive theoretical expectation.

There is a practical, less philosophically fraught dimension to the Casimir effect worth mentioning as well, because it illustrates how even the strangest corners of fundamental physics eventually intersect with ordinary engineering concerns. As technology has advanced toward building devices at ever-smaller scales, particularly in the field of micro-electro-mechanical systems, tiny mechanical devices, sensors, and actuators fabricated at scales measured in micrometers, engineers have found that the Casimir force, negligible at any scale humans typically interact with directly, becomes a genuinely significant, sometimes problematic factor at these tiny dimensions. Components positioned close enough together within a microscale device can experience Casimir attraction strong enough to cause unwanted sticking, a phenomenon engineers now have to explicitly account for and design around when building increasingly miniaturized technology, a strange but concrete example of quantum vacuum physics, first predicted as an abstract theoretical curiosity in 1948, having genuine practical consequences for cutting-edge engineering three quarters of a century later.

The Casimir effect also continues to serve modern physics research in a more exploratory capacity, functioning as one of the most sensitive available probes for testing whether gravity might behave differently than expected at extremely short distances, on the scale of microns or less, a question of real interest to physicists exploring theories involving hidden extra spatial dimensions or modifications to how gravity operates at very small scales. Because the Casimir force between two plates depends predictably on their exact separation, any unexpected deviation from that predicted dependence, measured with sufficient precision, could reveal the presence of new, currently unknown physics operating at short range, additional forces or dimensional effects that standard physics does not currently predict. As of this writing, no such deviation has been convincingly detected, but Casimir force measurements remain one of the active experimental frontiers where physicists continue to search, using nothing more exotic in principle than Casimir's original two metal plates, refined across seventy years of increasingly precise engineering.

It is worth noting, too, that Casimir was not working entirely alone when he made his breakthrough calculation, and the collaborative texture of that discovery is itself instructive. Casimir developed the closely related Casimir-Polder force together with his colleague Dirk Polder in 1947, describing the more general case of the force acting between a single neutral atom and a nearby conducting surface, before Casimir extracted from that broader analysis the simpler, more experimentally tractable case of two parallel plates that has since become the effect's standard textbook illustration. This detail matters because it illustrates something true across nearly every discovery examined in this book: even results now attributed to a single named scientist typically emerge from an ongoing conversation among collaborators and colleagues, incremental insights building on each other, rather than a single isolated flash of individual genius. Casimir's own account of the discovery credits a conversation with Niels Bohr, one of the founding figures of quantum mechanics, with nudging him toward framing the effect specifically in terms of zero point vacuum energy rather than the more roundabout van der Waals force calculations he and Polder had originally been pursuing, a reminder that even foundational physics breakthroughs often owe something to the right conversation happening at the right moment.

Hendrik Casimir lived to see his prediction experimentally confirmed, though only barely; Lamoreaux's definitive 1997 measurement came just a few years before Casimir's death in 2000, closing out a nearly fifty-year arc between theoretical prediction and experimental confirmation that stands as a genuine testament to both the predictive power of quantum field theory and the decades of patient engineering refinement it can take to actually put such predictions to the test.

It is worth stepping back, at the close of this chapter, to notice how much conceptual ground has been covered across the first eight chapters of this book, because the connections between these ideas are easy to lose amid the individual details of each discovery. Dark matter, examined across chapters two through five, revealed that the majority of the universe's matter is invisible, detectable only through its gravitational influence on the visible stars and galaxies it surrounds. Dark energy, the subject of chapters six and seven, revealed that empty space itself carries an intrinsic energy, one whose repulsive gravitational effect is currently overpowering the combined attractive gravity of every galaxy, star, and dark matter halo in the observable universe, driving cosmic expansion to accelerate rather than decelerate. This chapter, examining the quantum vacuum and the Casimir effect, has shown that the same basic notion, energy residing in seemingly empty space, is not merely a cosmological abstraction invoked to explain a puzzling supernova measurement, but a directly measurable laboratory phenomenon, confirmed to sub-percent precision using nothing more exotic than two metal plates and sufficiently careful engineering. Three chapters, three very different scales of investigation, one recurring theme: what looks empty, whether it is the space between galaxy clusters or the space between two plates in a laboratory vacuum chamber, is very often not empty at all.

The remaining four chapters of this book's investigation into the four great invisible phenomena turn to a different kind of nothing entirely: not matter too dim to see, not energy inherent to empty space, but ripples in the very geometry of space and time, produced by some of the most violent events the universe has ever produced, and detected, after a hundred year wait, by an instrument capable of noticing a change in distance thousands of times smaller than the width of a single proton. In 1916, Albert Einstein predicted the existence of these ripples, a direct mathematical consequence of his general theory of relativity, waves capable of carrying real energy across the universe despite being made of nothing more than the geometry of space and time distorting as they pass. Einstein himself doubted these ripples, called gravitational waves, would ever be strong enough for any conceivable instrument to detect. It would take almost exactly a century, and an instrument of engineering ambition that would have seemed like science fiction in 1916, to prove him both right about their existence and wrong about their detectability. That story begins in the next chapter.

CHAPTER NINE
Einstein's Ripples

There is a particular kind of scientific irony worth savoring before this chapter properly begins: the man who predicted gravitational waves in 1916 did not, for a stretch of time afterward, actually believe they were real.

To understand how Einstein arrived at gravitational waves at all, and why his confidence in them wavered so strangely even after he had derived them mathematically, you need some grounding in the theory that produced the prediction in the first place: general relativity, published in its final form in November 1915, arguably the single most successful and far-reaching theoretical achievement in the history of physics.

General relativity begins with a deceptively simple insight, one Einstein later described as the happiest thought of his life. Imagine a person in freefall, falling from a great height with nothing to slow their descent. Until they hit the ground, they would feel weightless, exactly the same sensation astronauts experience in orbit. Einstein realized that this was not a coincidence or an approximation, but a genuinely deep clue about the nature of gravity itself: a freely falling observer experiences no local sensation of gravitational force at all, which suggested that gravity might not be a force in the traditional sense, a pull acting between two objects across a distance, the way Isaac Newton had described it for over two centuries. Instead, Einstein began to suspect that gravity might be better understood as a consequence of geometry: the shape of space and time itself.

This idea, developed over the following decade with considerable mathematical difficulty, matured into general relativity's central claim: mass and energy curve the fabric of spacetime, a four-dimensional combination of the three familiar spatial dimensions plus time, and objects moving through curved spacetime follow paths that, from a local perspective, look exactly like what we call gravitational attraction, without requiring any force to be transmitted directly between the objects involved. The often-used analogy, imperfect but genuinely useful, pictures spacetime as a stretched rubber sheet, with a massive object like the Sun creating a deep depression in that sheet, and smaller objects, like planets, following curved paths around that depression not because the Sun is pulling on them directly, but because they are simply following the straightest possible path available within a space that has been warped by the Sun's presence. Remove the depression, and a planet would travel in a straight line. With the depression there, that straight line, in curved spacetime, becomes what we perceive as an orbit.

This was a genuinely radical reconceptualization of gravity, and it needed to prove itself against real, testable predictions before the physics community would accept it over Newton's far simpler, far more established framework, a framework that had, after all, successfully predicted the existence of the planet Neptune decades earlier, purely from analyzing irregularities in the orbit of Uranus, and that continued to work with essentially perfect accuracy for nearly every ordinary gravitational calculation scientists needed to perform. Overturning, or more precisely extending, a theory with that kind of track record required more than mathematical elegance. It required specific, falsifiable predictions that differed measurably from what Newton's older framework would predict, tested against real astronomical observations where the two theories genuinely disagreed. General relativity's first major triumph came almost immediately: the theory correctly predicted a subtle anomaly in the orbit of Mercury, the innermost planet, whose orbit precesses (gradually rotates around the Sun over time) at a rate slightly different from what Newtonian physics alone could explain, a discrepancy astronomers had puzzled over for decades. Einstein's equations accounted for the anomaly precisely, without requiring any additional adjustable parameters, a deeply satisfying resolution to a long-standing puzzle. An even more dramatic confirmation followed in 1919, when the British astronomer Arthur Eddington led an expedition to observe a total solar eclipse, timing the observation specifically to measure whether starlight passing near the Sun's edge would be bent by the Sun's gravitational field by the precise amount general relativity predicted, a bending too subtle to observe under normal daylight conditions but detectable during the brief darkness of a total eclipse, when stars near the Sun's position become momentarily visible. Eddington's measurements matched Einstein's prediction, and the announcement made Einstein, almost overnight, into the most famous scientist in the world, his name becoming a byword for genius in a way that persists to this day.

Amid all of this theoretical development, in 1916, Einstein turned his attention to a further mathematical consequence of his own equations, one that would take an entire century to test directly. If mass and energy curve spacetime, and if that curvature is what produces the effects we perceive as gravity, then it followed logically that a change in the distribution of mass, an object accelerating, or two massive objects orbiting each other, should produce a corresponding change, a ripple, in the surrounding curvature of spacetime, propagating outward from the source in a wave, much the way a stone dropped into a pond produces ripples spreading across the water's surface. Einstein worked through the mathematics and found that his equations did indeed predict such waves, and further, that these waves should propagate outward at a very specific speed: the speed of light, precisely 299,792 kilometers per second, the same universal speed limit that appears throughout the rest of relativity theory.

Here is where the story takes its curious turn. Despite having derived gravitational waves as a direct mathematical consequence of his own theory,

Einstein's private correspondence from this same period reveals genuine skepticism about whether they were physically real, as opposed to a mathematical artifact of the particular coordinate system used in the calculation, essentially an illusion produced by the specific mathematical bookkeeping method rather than a description of any genuine physical phenomenon. In February 1916, responding to a letter from the astronomer Karl Schwarzschild, Einstein wrote plainly that there are no gravitational waves analogous to light waves, a striking statement given that Einstein himself would, within a matter of months, publish his own paper formally introducing the concept. Even in that 1916 paper, titled Approximate Integration of the Field Equations of Gravitation, Einstein offered no particular confidence that such waves would ever actually be detected, treating the prediction as an interesting mathematical consequence of the theory rather than a phenomenon he expected astronomers to eventually observe directly.

This uncertainty persisted for Einstein for years, and in 1936, two decades after his original prediction, Einstein and a collaborator, Nathan Rosen, submitted a paper to the journal Physical Review that initially argued, based on a reexamination of the underlying mathematics, that gravitational waves might not exist as genuine physical phenomena after all, a dramatic reversal of his earlier position. That paper, as it happened, contained a technical error in its mathematical setup, caught during peer review, and

Einstein, initially resistant to the criticism, eventually revised his conclusion after further work with a different collaborator, Leopold Infeld, ultimately reaffirming that gravitational waves were indeed a genuine prediction of the theory. This episode is worth including here not to diminish Einstein's towering achievement, but because it offers a genuinely useful, humanizing corrective to the popular image of Einstein as an infallible oracle whose every pronouncement arrived polished and certain. Even the theory's own creator wrestled, across decades, with whether one of its most striking predictions was physically real, working through genuine confusion, genuine error, and genuine revision before arriving at the settled understanding that eventually became standard physics.

It is worth spending a moment on why gravitational waves, once accepted as theoretically real, were expected to be so fantastically difficult to detect, because this expectation shaped nearly a century of subsequent research and explains why direct detection took until 2015 to achieve. The strength of a gravitational wave, as it passes through a given region of space, is described by a quantity called strain, essentially the fractional change in distance between two points caused by the wave's passage. For gravitational waves produced by anything happening here on Earth, or even within our solar system, this strain is so vanishingly small as to be utterly undetectable by any conceivable instrument. Even the most violent events astrophysics can produce, two black holes spiraling into each other and merging, releasing more energy in a fraction of a second than every star in the observable universe combined radiates in that same instant, produce a strain, by the time that wave has traveled the vast distance to Earth, of roughly one part in ten to the twenty first power. To put a number that abstract into more graspable terms: detecting a gravitational wave of that strength requires measuring a change in distance, across a baseline of several kilometers, smaller than one ten-thousandth the width of a single proton, a level of precision that sounds, on first encounter, like a physical impossibility rather than an engineering challenge.

For most of the twentieth century, this was, practically speaking, exactly what it seemed: an all but impossible measurement problem, sufficiently daunting that most physicists regarded direct detection as a worthy long-term goal but not a realistic near-term prospect. Indirect evidence for gravitational waves did eventually arrive well before any direct detection, and this indirect confirmation deserves its own recognition, because it represents an important, often underappreciated milestone in its own right. In 1974, the astronomers Russell Hulse and Joseph Taylor discovered a binary pulsar system, two extremely dense, rapidly rotating neutron stars orbiting each other, one of which periodically sweeps a beam of radio emission across Earth like a cosmic lighthouse, allowing extraordinarily precise timing measurements of the system's orbital period. General relativity predicts that such a system should lose orbital energy over time, gradually, through the emission of gravitational waves, causing the two neutron stars to spiral very slowly closer together and their orbital period to correspondingly shorten. Hulse and Taylor tracked this system's orbital decay over years of careful observation, and found it matched general relativity's predicted rate of energy loss due to gravitational wave emission with remarkable precision, an indirect but genuinely compelling confirmation that gravitational waves, whatever their ultimate detectability, were carrying away real energy exactly as Einstein's theory predicted. This discovery earned Hulse and Taylor the Nobel Prize in Physics in 1993, a recognition of what remains, to this day, one of the most elegant indirect tests of general relativity ever conducted.

But indirect confirmation, however elegant, is not the same as direct detection, and the difference matters both scientifically and, frankly, dramatically. An indirect measurement confirms that a predicted effect is consistent with observed consequences; a direct detection means actually capturing the wave itself, passing through an instrument, in real time, carrying with it genuine information about the specific astrophysical event that produced it. By the 1970s and 1980s, a small community of physicists had begun seriously contemplating what kind of instrument could possibly achieve such a feat, and the answer that eventually emerged, after decades of technological development, false starts, and sustained institutional persistence against considerable scientific skepticism, would become one of the most ambitious and successful large-scale scientific instruments ever constructed.

Before turning to that instrument, it is worth returning briefly to Eddington's 1919 eclipse expedition, because the story deserves a slightly fuller telling than a single paragraph can provide, and because it illustrates something important about how extraordinary scientific claims get tested against the sky itself rather than settled by argument alone. Eddington, working with a team that split its observing effort between the island of Principe off the coast of West Africa and a site in Sobral, Brazil, faced considerable practical difficulty: clouds threatened to obscure the observation entirely at Principe, and the team had only the brief window of totality, a few minutes at most, in which to photograph the positions of stars near the Sun's darkened disk before comparing those positions against photographs of the same stars taken months earlier, when the Sun was nowhere near that part of the sky. The comparison would reveal whether starlight passing close to the Sun had been deflected by the precise amount, twice the deflection Newtonian physics alone would predict, that Einstein's general relativity called for. When Eddington's results, though not without some measurement uncertainty that later reanalysis has scrutinized carefully, came down clearly on the side of Einstein's prediction rather than the Newtonian one, the announcement, made jointly to the Royal

Society and the Royal Astronomical Society in London in November 1919, was reported around the world, propelling Einstein to a form of global celebrity that few scientists before or since have experienced. It is worth noting, in fairness to the historical record, that subsequent, more rigorous reanalysis of the original 1919 data and methodology by later historians of science has found the original measurements to have been less unambiguous than the triumphant contemporary press coverage suggested, though subsequent, far more precise tests of light bending conducted throughout the twentieth century, using radio astronomy and other refined techniques, have overwhelmingly confirmed Einstein's prediction to extraordinary precision, removing any serious lingering doubt about the underlying physics, even if the original 1919 measurement was, in hindsight, closer to the edge of its own experimental capability than the celebratory headlines of the time let on.

The broader pattern here, a bold theoretical prediction, initial confirmation that captures public imagination, followed by decades of increasingly rigorous refinement and testing, will recur throughout the remainder of this book, and it is worth having named clearly at this stage: single confirming observations, however dramatic, are rarely the end of a scientific story. They are typically the beginning of a much longer process of independent replication, methodological scrutiny, and cumulative refinement, a process that took general relativity itself the better part of a century to complete, and that would prove, if anything, even more demanding for the specific prediction of gravitational waves, which is where the next chapter picks up the story.

CHAPTER TEN

The Century Long Wait

Rainer Weiss did not set out to build the most sensitive scientific instrument in human history. In 1967, he was a young physicist at MIT, assigned to teach a course on general relativity despite having, by his own later admission, only a shaky grasp of the subject himself. Forced to think carefully through the theory in order to explain it to students, Weiss found himself puzzling over a practical question that had mostly been treated, up to that point, as an abstract theoretical curiosity: if gravitational waves were real, as Einstein's mathematics predicted, how would you actually go about detecting one?

The dominant approach at the time, developed by the physicist Joseph Weber at the University of Maryland starting in the 1960s, used large aluminum cylinders called resonant bar detectors, instruments designed to physically vibrate, ever so slightly, when a passing gravitational wave stretched and compressed them. Weber announced, in the late 1960s and early 1970s, that his detectors had recorded signals consistent with gravitational waves, results that generated enormous excitement, briefly suggesting that direct detection might already be within reach decades earlier than anyone had dared hope. Unfortunately, despite Weber's genuine conviction and considerable effort, other laboratories around the world, building similar bar detectors specifically to replicate his results, could not confirm his findings, and the broader physics community gradually concluded, after years of careful, sometimes contentious scrutiny, that Weber's reported detections were not genuine gravitational wave signals, but most likely statistical artifacts or instrumental noise misinterpreted as a real astrophysical discovery. It is worth including this episode honestly, rather than skipping past it, because it illustrates something genuinely important about how science actually polices itself: a claimed discovery, however sincerely believed by its discoverer, has to survive independent replication by other research groups before the broader community accepts it as real, and Weber's bar detector results, however historically significant in galvanizing interest in the field, ultimately did not survive that test.

Weiss took a fundamentally different technical approach, one he first worked out in detail in a 1972 paper written for MIT's Research Laboratory of Electronics, a document Weiss himself, in a moment of characteristic understatement, initially titled using the term antenna rather than the more familiar language of detector or observatory that would later become standard. Rather than measuring the vibration of a solid metal bar, Weiss proposed using a laser interferometer: an instrument that splits a single laser beam into two perpendicular paths, sends each half down a long arm to a mirror positioned at the far end, reflects both beams back to their starting point, and then recombines them. Under normal, undisturbed conditions, the two beams, having traveled identical distances, recombine in a precisely predictable way, producing a stable, unchanging interference pattern. But if a gravitational wave passes through the instrument while the light is in transit, it should, according to general relativity, stretch one arm of the interferometer by an infinitesimal amount while simultaneously compressing the perpendicular arm, a signature effect of how gravitational waves distort spacetime, altering the relative path lengths the two light beams travel and producing a correspondingly tiny, but in principle detectable, shift in the recombined interference pattern.

This was, on paper, a beautifully elegant solution to a fantastically difficult measurement problem. Turning it into an actual working instrument required overcoming engineering obstacles substantial enough that the project would consume, in total, more than four decades of sustained effort before finally succeeding. Weiss's 1972 blueprint attracted the attention of Kip Thorne, a Caltech theoretical physicist who had, since 1968, led his own research group studying the theory of gravitational waves and their likely astrophysical sources, without, at that stage, a serious experimental program to complement the theoretical work. The two met in 1975, and Thorne, persuaded by Weiss's case that a sufficiently large interferometer had a genuine, realistic chance of achieving direct detection, began championing the project at Caltech, eventually recruiting the Scottish physicist Ronald Drever from the University of Glasgow in 1979 to lead the experimental side of Caltech's effort, since Drever had already established himself as one of the leading builders of prototype gravitational wave interferometers in Europe.

Through the late 1970s and into the 1980s, both the Caltech group under Drever and Weiss's own team at MIT built successively larger prototype interferometers, proving out the underlying concept at small scale, a forty meter prototype at Caltech and a smaller one point five meter prototype at MIT, while simultaneously identifying and solving an accumulating list of technical problems that stood between a promising laboratory demonstration and an instrument actually sensitive enough to detect a real astrophysical gravitational wave. Seismic vibration, from sources as mundane as distant traffic, ocean waves crashing against faraway coastlines, or even the footsteps of researchers walking through the building, had to be isolated from the sensitive optical components through elaborate suspension systems. Laser stability had to be improved dramatically, since any fluctuation in the laser's own frequency could mimic or mask the signal researchers were hunting for, a problem Drever helped solve through a laser stabilization technique now known as the Pound-Drever-Hall technique, still used throughout precision optics today, well beyond gravitational wave research. And the entire apparatus needed to operate within an ultra-high vacuum, since even ordinary air molecules, colliding randomly with the system's mirrors, would introduce enough noise to overwhelm the exquisitely small signal a genuine gravitational wave would produce.

By 1984, the case for building a full-scale instrument had matured enough that Caltech and MIT formally signed an agreement to jointly design and construct what would become known as the Laser Interferometer Gravitational-Wave Observatory, LIGO, with joint leadership initially shared among Drever, Weiss, and Thorne, a partnership that, like many ambitious scientific collaborations involving strong personalities and genuine scientific disagreement about technical direction, was not always smooth, but that nonetheless carried the project through a rigorous, multi-year review by the National Science Foundation, LIGO's primary funding agency, culminating in a weeklong, in-depth technical evaluation by an external review committee in November 1986. The National Science Foundation ultimately committed to funding LIGO's construction, and by 1990, the project received the largest single grant the foundation had ever awarded up to that point, a reflection of both the scientific ambition involved and the sheer scale of engineering required to build not one but two identical, enormous interferometers, positioned deliberately at separated locations, one in Hanford, in the high desert of eastern Washington State, and the other in

Livingston, Louisiana, roughly three thousand kilometers apart.

The choice to build two separated detectors, rather than one, reflected a hard-won lesson from Weber's earlier, ultimately discredited bar detector results: a genuine gravitational wave signal should arrive at widely separated detectors within a predictable, extremely short time window, determined by the speed of light and the distance between the two sites, while local sources of noise, seismic disturbances, electronic glitches, or other spurious signals, would generally not produce matching signals at both distant locations simultaneously. Requiring coincident detection at two independent, separated sites became, and remains, one of LIGO's most powerful tools for distinguishing a genuine astrophysical signal from the enormous background of terrestrial noise any sufficiently sensitive instrument inevitably picks up.

Each LIGO detector took the shape of an L, with two perpendicular arms extending four kilometers in each direction, vacuum tubes roughly four feet in diameter running the entire length of each arm, evacuated to a vacuum purer than the vacuum of interplanetary space, housing the laser beam path and requiring a level of large-scale precision engineering that had genuinely never been attempted before at this particular combination of scale and sensitivity. Construction proceeded through the 1990s, and the initial version of LIGO, now referred to as Initial LIGO to distinguish it from later, more sensitive upgrades, began its first observational data-taking runs in 2002. For roughly the next eight years, Initial LIGO operated, along with its European counterpart Virgo, a broadly similar interferometer built through a collaboration of French, Italian, Dutch, Polish, and Hungarian research institutions near Pisa, Italy, searching for gravitational wave signals across a series of observing runs. None were found.

This might sound, on its face, like eight years of failure, but it represented something closer to essential, expected groundwork. Initial LIGO was never expected to be sensitive enough to guarantee a detection; it was, from the outset, conceived as a proof of concept and engineering testbed, intended to demonstrate that the interferometer approach could actually function at the required scale and sensitivity, uncovering and resolving the countless additional sources of noise and instability that only become apparent when operating a real, full-scale instrument continuously over extended periods, problems that no amount of prototype testing at smaller scale could have fully anticipated. Throughout this period, the LIGO collaboration, formally organized as the LIGO Scientific Collaboration starting in 1997 under the leadership of the physicist Barry Barish, who took over as the project's director and is credited with transforming LIGO from an ambitious but somewhat loosely organized university partnership into a large, disciplined, internationally collaborative scientific enterprise, was simultaneously planning and developing a far more sensitive successor instrument, Advanced LIGO, incorporating years of accumulated lessons about noise reduction, improved mirror coatings, more powerful and stable lasers, and refined suspension systems, all aimed at boosting the detector's sensitivity by a factor of roughly ten compared to the initial configuration, translating into a thousandfold increase in the volume of space the detector could effectively search for gravitational wave sources, since detection volume scales with the cube of sensitivity.

Advanced LIGO's construction and installation proceeded through the early 2010s, with the upgraded detectors gradually brought online and progressively tuned toward their full design sensitivity. In September 2015, with Advanced LIGO still in an early engineering test run, not yet formally collecting official science data, and with sensitivity still well short of its ultimate design target, something arrived that nobody on the collaboration was fully expecting so soon.

It is worth pausing here to reflect on the sheer duration of this undertaking, because the number is easy to state and harder to truly absorb. Weiss first sketched the interferometer concept in 1972. Direct detection would not come until 2015, a gap of forty three years, spanning the bulk of an entire scientific career, during which the project survived skeptical funding reviews, competing technical approaches from rival research groups, internal leadership disputes serious enough to require external mediation, and, for a significant portion of that period, sincere scientific doubt among some physicists outside the collaboration about whether interferometric detection was achievable at all with technology available in this generation. Weiss, Thorne, and Drever were, by the time success finally arrived, no longer young men; Weiss was eighty three years old in 2015, having devoted essentially his entire working scientific life to an instrument whose success remained genuinely uncertain for the overwhelming majority of that time. This is worth naming plainly because popular accounts of scientific discovery often compress this kind of multi-decade institutional persistence into a tidy origin story, a flash of insight followed swiftly by triumphant confirmation. The real history of LIGO looks nothing like that. It looks like four decades of unglamorous, incremental engineering work, sustained institutional fundraising, and a research community's collective decision, renewed again and again across multiple funding cycles, that a currently unproven technology was worth continuing to pursue, based on theoretical confidence in the underlying physics and steady, if slow, experimental progress toward the sensitivity such physics demanded.

There is also a broader institutional lesson embedded in LIGO's long gestation, one relevant well beyond gravitational wave physics specifically. Large scientific instruments of this kind, whether LIGO, the underground dark matter detectors discussed earlier in this book, or the particle accelerators used to study fundamental physics at the smallest scales, require a particular and somewhat unusual combination of institutional patience: funding agencies willing to commit sustained resources across decades without guaranteed results, university administrations willing to house speculative, expensive projects for years before any scientific payoff materializes, and individual scientists willing to build careers around instruments that might, for entirely legitimate reasons rooted in the underlying physics rather than any failure of effort, simply never succeed within their working lifetimes. Rainer Weiss has spoken candidly, in interviews given well after LIGO's eventual success, about genuinely not knowing, for long stretches of the project's development, whether detection would ultimately prove possible with technology available during his career, a level of honest uncertainty that makes the eventual triumph land with considerably more weight than a story of confident, foreordained success ever could.

It is also worth noting how LIGO's technical requirements pushed forward entire adjacent fields of engineering and materials science, independent of whether gravitational waves were ever successfully detected. The mirror coatings developed for LIGO's optical components, required to minimize thermal noise and optical absorption to a degree far beyond any prior commercial or scientific application, advanced the broader field of precision optical coating technology. The vacuum systems built for LIGO's four kilometer arms remain, to this day, among the largest and most sophisticated ultra-high vacuum systems ever constructed for any purpose, pushing forward vacuum engineering techniques with applications well beyond gravitational wave astronomy. The seismic isolation systems developed to shield LIGO's mirrors from ground vibration have influenced precision engineering in fields ranging from semiconductor manufacturing to other sensitive physics experiments entirely unrelated to gravitational waves. This is a pattern worth naming because it recurs throughout the history of large, ambitious scientific instruments: the technological spillover from attempting something at the absolute edge of what is currently possible often provides lasting value independent of whether the primary scientific goal is ultimately achieved, a kind of built-in insurance against the very real possibility, present throughout LIGO's four-decade development, that the central prediction being tested might never actually be confirmed.

The signal that arrived in September 2015, and the extraordinary days of verification, disbelief, and quiet, careful confirmation that followed its arrival, form the subject of the next chapter, one of the genuine landmark moments in the history of experimental physics, and the moment that finally, after that four-decade wait, transformed gravitational waves from a mathematically elegant prediction into a directly observed physical reality.

CHAPTER ELEVEN

The Day Spacetime Chirped

At 11:53 in the morning, local time, on September 14, 2015, an automated email landed in the inbox of a thirty three year old postdoctoral physicist named Marco Drago. Drago was not in Louisiana or Washington State, where LIGO's two massive detectors physically sit. He was in Hanover, Germany, at the Max Planck Institute for Gravitational Physics, sitting at his desk, on the phone with a colleague in Italy, running one of four independent automated data analysis pipelines that continuously combed through the raw output streaming from both LIGO detectors, hunting for anything that might resemble a genuine gravitational wave signal buried within the ordinary noise. Drago received roughly one such alert email per day, almost always false alarms, statistical blips in the data that resembled a signal only superficially before further analysis ruled them out.

This one was different, and the difference was immediately, unmistakably obvious. The email contained links to two plots, one from each LIGO detector, and both showed the same distinctive shape: a signal that swept rapidly upward in frequency and amplitude before abruptly cutting off, a pattern gravitational wave physicists had spent decades theoretically predicting and had informally nicknamed a chirp, because, when converted into an audible sound wave, it resembles exactly that, a brief, rising chirp, not unlike a bird's call compressed into a fraction of a second. The signal's strength was extraordinary. Measured by a statistic called signal to noise ratio, essentially a measure of how clearly a signal stands out above the instrument's ordinary background noise, this event registered at 24, more than double the value of 10 that researchers typically considered the minimum threshold worth taking seriously. Drago's own reaction, by his later account, was disbelief, not excitement, because a signal this strong and this clean seemed, if anything, too good to be a genuine first detection.

There was a specific, well-founded reason for that skepticism, one built directly into how the LIGO collaboration had structured its entire detection and verification process. To guard against the very real danger of researchers unconsciously wanting to see a signal so badly that they might convince themselves a piece of noise was real, LIGO maintained a practice called blind injection, in which a small, carefully restricted group of senior scientists retained the ability to secretly insert a fake, artificially generated signal into the detector's data stream, without informing the rest of the collaboration, specifically to test whether the analysis pipelines and the human researchers running them would correctly identify and validate a genuine-looking signal through the full verification process. Nobody analyzing the September 14 data initially knew, and could not know for certain until an official check was performed, whether what Drago and, within minutes, dozens of other LIGO scientists around the world were looking at was a real astrophysical event or an internally planted test, a blind injection deliberately designed to be indistinguishable from the real thing until the very final stage of review.

What followed, over the coming days and weeks, was one of the most rigorous verification processes in the history of experimental physics, precisely because the stakes of getting it wrong, either falsely claiming a monumental discovery or, just as damaging to the field's credibility, dismissing a genuine detection out of excessive caution, were so enormous. The LIGO Scientific Collaboration, by this point numbering more than a thousand scientists across dozens of institutions worldwide, mobilized an extensive, methodical investigation. Researchers checked whether the signal could plausibly have been produced by any known source of instrumental or environmental noise: seismic activity, electromagnetic interference, a stray truck driving past one of the remote detector sites, a software glitch in the data acquisition system, or any of the roughly one hundred thousand auxiliary data channels LIGO continuously monitors precisely to catch and rule out exactly this kind of contaminating signal. None of these explanations held up. The signal appeared in both detectors, at Hanford, Washington, and Livingston, Louisiana, separated by roughly three thousand kilometers, arriving at Livingston approximately seven milliseconds before it arrived at

Hanford, a time delay entirely consistent with a gravitational wave traveling at the speed of light and reaching the two detectors from a particular direction in the sky, and thoroughly inconsistent with any plausible local, terrestrial source of interference that could somehow produce matching signals at two such widely separated locations.

Crucially, and this detail matters enormously for understanding why the eventual announcement carried such scientific weight, the shape of the observed signal did not merely look vaguely like a gravitational wave in some general sense. It matched, with remarkable precision, the specific waveform predicted by numerical simulations of two black holes spiraling into each other and merging, calculations that themselves represented decades of dedicated theoretical and computational effort to solve Einstein's famously difficult field equations for exactly this kind of extreme, highly dynamic gravitational scenario. By comparing the observed signal against an enormous library of pre-computed template waveforms, each corresponding to a different possible combination of black hole masses and orbital parameters, researchers were able to determine not just that the signal was consistent with a black hole merger in general, but to extract specific physical properties of the actual system that had produced it.

The numbers that emerged from this analysis were themselves staggering. The signal, formally designated GW150914 following LIGO's naming convention of combining the letters GW for gravitational wave with the date of detection, originated from the merger of two black holes with masses of approximately 36 and 29 times the mass of our Sun, spiraling together and merging into a single black hole of approximately 62 solar masses. Careful readers will notice that thirty six plus twenty nine does not equal sixty two, and this discrepancy is not an error. The missing three solar masses, an amount of mass roughly equivalent to three entire Suns, had been converted directly into energy and radiated away as gravitational waves during the collision itself, released in a final cataclysmic fraction of a second, an amount of energy that briefly made this single, otherwise unremarkable collision between two black holes, unremarkable in the sense that such mergers likely occur regularly throughout the universe, radiate more power than the combined light output of every star in the observable universe put together, all of it carried not as light but as pure ripples in the geometry of spacetime itself.

The event had occurred, according to the analysis, at a distance of roughly 1.3 billion light years from Earth, meaning the actual collision took place 1.3 billion years ago, long before complex multicellular life had even emerged on our own planet, and the resulting ripple in spacetime had been traveling toward Earth, at the speed of light, for the entirety of that immense span, arriving at LIGO's two detectors within milliseconds of each other on an otherwise ordinary Monday morning in September 2015.

For nearly five months after that initial detection, the LIGO collaboration maintained strict internal confidentiality while completing its exhaustive verification process, a period that must have been almost unbearably difficult for the scientists directly involved, sitting on what they increasingly suspected was one of the most significant experimental physics results in a century, unable to share it publicly until every conceivable alternative explanation had been rigorously excluded and the analysis had cleared the full internal review process, including formal confirmation that the signal had not, in fact, been a blind injection test, a confirmation that came only after the relevant safe holding the injection controls was verified as empty. The collaboration also used this period to prepare a companion scientific paper analyzing the astrophysical implications of the detection in detail, and to coordinate simultaneous announcement plans across the international collaboration, since a discovery of this magnitude, involving over a thousand collaborating scientists across multiple countries, required careful coordination to announce properly and credit appropriately.

On February 11, 2016, the LIGO Scientific Collaboration, together with the European Virgo collaboration, formally announced the detection at a press conference, publishing the full technical details simultaneously in a paper in the journal Physical Review Letters titled, with characteristic scientific understatement given the magnitude of what it described, Observation of Gravitational Waves from a Binary Black Hole Merger. The announcement made headlines around the world, and it is worth pausing to note just how large the resulting public response was: the discovery generated coverage across virtually every major news outlet globally, tens of millions of mentions across social media platforms within days, and traffic to the Physical Review Letters website sufficiently intense that the journal's servers reportedly struggled to keep pace with download requests in the hours immediately following the announcement.

The scientific significance extended well beyond the sheer scale of public interest. GW150914 represented, simultaneously, the first direct detection of gravitational waves in history, exactly one hundred years after Einstein's original 1916 theoretical prediction; the first direct observational confirmation that binary black hole systems, two black holes orbiting each other closely enough to eventually merge, actually exist in nature, something theoretical astrophysicists had long suspected but had never before directly observed; and the first direct observation of a black hole actually being formed, captured in the specific moment of the two progenitor black holes merging into the single, larger remnant. Few single measurements in the history of experimental physics have delivered so many separate, independently significant confirmations within one event.

Recognition followed swiftly, by the standards of Nobel Prize timelines. In 2017, barely eighteen months after the announcement, the Nobel Prize in Physics was awarded jointly to Rainer Weiss, Barry Barish, and Kip Thorne, cited specifically for what the Nobel committee's announcement described as decisive contributions to the LIGO detector and the observation of gravitational waves. Ronald Drever, whose engineering contributions had been every bit as essential to LIGO's eventual success as those of his two co-laureates, had died in March 2017, only months before the prize was announced; the Nobel Prize is never awarded posthumously, a rule that meant Drever's foundational role, while acknowledged repeatedly and warmly by his surviving colleagues in subsequent interviews and public statements, went unrecognized by the prize itself, a poignant coda to a project that had, quite literally, taken longer to succeed than many of its earliest architects had any confident right to expect they would personally live to see.

Marco Drago, the postdoc whose automated pipeline first flagged the signal on that September morning, would later describe the experience with a characteristic mixture of scientific caution and genuine wonder: it was immediately clear, he said, that he was looking at something extraordinary, even before the months of painstaking verification could formally confirm what his instincts, trained through years of careful pipeline development, were already telling him. That combination, immediate recognition tempered by rigorous, patient, months-long verification, captures something essential about how the most significant discoveries in physics actually happen: not as a single triumphant moment of certainty, but as a first flicker of genuine signal, followed by an enormous, careful, collective effort to make absolutely sure that flicker was real before telling the world.

It is worth explaining, in a bit more technical depth than the narrative above allowed, exactly how researchers extracted specific numbers, those 36 and 29 solar mass black holes, that 1.3 billion light year distance, from what was, in raw form, nothing more than a tiny, rapidly evolving wiggle in the position of a set of mirrors four kilometers apart. This process, called matched filtering, relies on a library of template waveforms, precomputed using numerical relativity, the specialized branch of computational physics dedicated to solving Einstein's field equations on supercomputers for scenarios too complex for exact analytical solutions, such as two massive, rapidly spinning black holes spiraling into one another. Researchers generate hundreds of thousands of these template waveforms in advance, each corresponding to a different possible combination of the two merging objects' masses, spins, and orbital orientations, essentially building an enormous catalog of predicted chirps, one for every plausible astrophysical scenario general relativity permits. When an actual signal like GW150914 arrives, an automated pipeline systematically compares the observed data against this entire template library, searching for the best possible match. The template that fits most closely, in this case corresponding to a merger of roughly 36 and 29 solar mass black holes, effectively tells researchers the physical properties of the source, translating a fleeting distortion in spacetime, lasting barely a fifth of a second from first detectable signal to final merger, into a detailed physical portrait of an event that occurred over a billion years ago and light years beyond anything a telescope could ever directly image.

This technique depends entirely on the extraordinary theoretical work, spanning decades, that went into solving Einstein's equations for the specific case of merging black holes, work that itself represents one of the great underappreciated triumphs of modern computational physics. Kip Thorne's research group, and numerous other numerical relativity groups worldwide, spent years developing the computational techniques necessary to simulate these mergers accurately, a problem that resisted straightforward solution for decades because the final, most violent moments of a black hole merger involve gravitational fields so strong and geometry so extremely curved that the simplified approximation techniques adequate for weaker gravitational fields simply break down, requiring instead the full, formidably complex machinery of Einstein's nonlinear field equations, solved numerically, one tiny time step at a time, on some of the most powerful supercomputers available to the physics community. Without this decades-long parallel investment in theoretical and computational astrophysics, LIGO's detectors could have recorded a perfectly genuine gravitational wave signal and had no reliable way to confirm what had produced it or extract meaningful physical information from the data, a reminder that GW150914's success was never purely an experimental achievement, but the product of two entirely separate, equally demanding scientific traditions, precision instrumentation and theoretical computation, converging at exactly the right moment.

The next chapter turns to what came after, the new field of gravitational wave astronomy that GW150914 inaugurated, and the remarkable catalog of cosmic collisions LIGO and its international partner observatories have detected in the years since that first, history-making September morning.

CHAPTER TWELVE

Listening to Black Holes Collide

GW150914 was, by design and by necessity, treated with extraordinary caution. Five months of verification preceded its public announcement, a level of scrutiny appropriate for an unprecedented, field-defining first detection. What is easy to forget, looking back from the present, is how quickly that caution gave way to a genuinely new kind of routine. Within four months of the first announcement, LIGO had already recorded a second confirmed detection, a merger of black holes with masses of roughly 14 and 8 solar masses, designated GW151226, detected on Boxing Day, December 26, 2015, during the same initial observing run that had produced GW150914. A third significant detection followed during LIGO's second observing run, another black hole merger, this one involving masses of roughly 31 and 19 solar masses. Gravitational wave astronomy, in other words, did not remain a field defined by a single extraordinary event for very long. It rapidly became an active, ongoing observational discipline, with new detections arriving at an accelerating pace as the LIGO and Virgo detectors were progressively refined and their sensitivity improved across successive observing runs.

The character of these early detections was consistent: binary black hole mergers, two black holes spiraling together and merging, producing gravitational wave signals but, because black holes by definition emit no light of any kind, nothing detectable through conventional telescopes. This made perfect sense given LIGO's underlying sensitivity at the time. Black hole mergers, involving extremely massive, extremely compact objects, produce the strongest gravitational wave signals of any known astrophysical source, making them the easiest targets for a still-maturing detector network to find. But astrophysicists had long anticipated that a different, in some ways even more scientifically exciting category of event should also be detectable: the merger of two neutron stars, the extraordinarily dense remnants left behind when massive stars explode as supernovae, objects so dense that a neutron star packs roughly the mass of our Sun into a sphere only about twelve miles across. Unlike black hole mergers, a neutron star merger was expected to produce not just gravitational waves but also a genuine burst of electromagnetic radiation, light in various forms, potentially detectable by conventional telescopes, offering the tantalizing prospect of observing the same cosmic event through two completely different, complementary channels simultaneously.

That prospect became reality on August 17, 2017, in an event that would come to be recognized as one of the single most scientifically productive nights in the history of observational astronomy. At 8:41 in the morning Eastern time, LIGO's two detectors, along with the European Virgo detector, which had recently joined the observing network and, crucially, helped narrow down the event's location in the sky through triangulation between three separated instruments rather than just two, recorded a gravitational wave signal with a distinctly different character from the black hole mergers detected previously: a longer, more gradual chirp, consistent with the merger of two much less massive, more slowly evolving objects, exactly the signature theorists had predicted for colliding neutron stars. This event, designated GW170817, triggered something LIGO scientists had spent years preparing for but had not yet had the opportunity to actually execute: an immediate, urgent alert broadcast to telescopes and observatories around the world, informing astronomers that a promising gravitational wave source had just been detected and providing a rough estimate of where in the sky to look.

Remarkably, just 1.7 seconds after the gravitational wave signal itself, NASA's Fermi Gamma-ray Space Telescope independently detected a short gamma ray burst, a brief, intense flash of high-energy radiation, arriving from a region of sky consistent with the LIGO-Virgo localization. This near-simultaneous detection, a gravitational wave signal followed within seconds by a gamma ray burst from the same patch of sky, provided immediate, powerful evidence that the two signals shared a common origin, and it confirmed a longstanding theoretical suspicion that at least some short gamma ray bursts, a previously mysterious category of astronomical event, are produced by exactly this kind of neutron star merger.

What followed over the subsequent hours, days, and weeks represented an extraordinary, coordinated global observational campaign. Less than eleven hours after the initial detection, the Swope Telescope in Chile identified a new, previously unseen point of light in a galaxy called NGC 4993, roughly 130 million light years from Earth, a discovery rapidly confirmed and followed up by dozens of additional observatories around the world, spanning the full range of the electromagnetic spectrum, from radio waves through infrared, visible light, ultraviolet, and X-rays. This newly discovered transient object, designated AT2017gfo, represented the first confirmed direct observation of a phenomenon theorists had predicted decades earlier but had never actually witnessed: a kilonova, an explosive event produced by neutron star mergers, distinct from the more familiar supernovae produced by dying stars, and powered by the radioactive decay of extraordinarily heavy elements freshly synthesized in the merger's aftermath.

This last detail carries genuine, far-reaching scientific significance that extends well beyond gravitational wave physics itself, into one of the oldest open questions in astrophysics: where do heavy elements, elements heavier than iron, including gold, platinum, and uranium, actually come from? Ordinary stellar nuclear fusion, the process powering stars throughout their normal lives, can build up elements only as heavy as iron before the fusion process stops releasing net energy, meaning heavier elements require some other, more extreme production mechanism. Physicists had long theorized that a specific nuclear process, called rapid neutron capture or the r-process, occurring in an environment extraordinarily rich in free neutrons, could synthesize these heavier elements, and neutron star mergers, involving literally an entire star's worth of neutron-rich material suddenly disrupted and ejected into space, had been proposed as one of the most promising candidate sites where this process might actually occur in nature. The detailed spectroscopic observations of AT2017gfo's kilonova, gathered across the following weeks by observatories worldwide, provided the first direct observational confirmation that neutron star mergers do indeed produce heavy elements through exactly this r-process mechanism, offering genuine, hard observational evidence for where a meaningful fraction of the universe's gold, platinum, and other heavy elements actually originated: not in the hearts of ordinary stars, but in the cataclysmic collision of two neutron stars, an event so energetic it briefly outshone the combined light of a billion ordinary stars while forging, in that same brief flash, precious metals that would eventually find their way into planets, asteroids, and, conceivably, into jewelry worn by observers on a planet not yet formed at the time of the collision.

GW170817 earned the designation multi-messenger astronomy, a term describing the coordinated study of a single astrophysical event using multiple, fundamentally different types of signal, in this case gravitational waves alongside electromagnetic radiation across nearly the entire observable spectrum. It represented, and remains as of this writing, the only confirmed instance where both a gravitational wave signal and a clear electromagnetic counterpart have been jointly observed and confirmed, a scarcity explained partly by the relative rarity of neutron star mergers compared to black hole mergers, and partly by the practical difficulty of successfully localizing and following up such events quickly enough, and with sufficiently favorable viewing geometry, to capture their brief electromagnetic afterglow before it fades.

The pace of gravitational wave detection has continued accelerating in the years since, as successive observing runs have brought progressively more sensitive detector configurations online, and as the international detector network has expanded to include not just LIGO's two American detectors and the European Virgo detector, but also KAGRA, an underground detector built in Japan using a somewhat different technical approach involving cryogenically cooled mirrors to further reduce thermal noise. By the end of the LIGO-Virgo-KAGRA collaboration's third major observing run, the combined catalog of confirmed gravitational wave detections had grown to roughly ninety separate events, the overwhelming majority representing binary black hole mergers, with a small additional number of confirmed or likely neutron star mergers and, intriguingly, several events consistent with the merger of a neutron star and a black hole, a third category of compact object collision that theorists had predicted but that had not been directly observed prior to the gravitational wave era. Subsequent observing runs have continued adding to this catalog at an increasing rate, and current projections from the collaboration anticipate that once next-generation detector upgrades and entirely new instruments, including a proposed detector called the Einstein Telescope in Europe and another called Cosmic Explorer in the United States, come online in the coming years, the annual detection rate for neutron star mergers alone could climb into the range of tens to potentially hundreds of events per year, transforming what began as a single, historic detection in 2015 into a routine, high-volume observational astronomy discipline, not unlike how optical astronomy transitioned, over the preceding century, from rare, celebrated individual discoveries to the systematic, large-scale survey science that characterizes much of the field today.

It is worth pausing, at the close of this chapter, to appreciate just how quickly gravitational wave astronomy has matured, because the speed of that transition is itself remarkable. It took physicists a full century to move from Einstein's theoretical prediction to the first confirmed detection. It has taken barely a decade to move from that first detection to a mature, internationally coordinated observational discipline generating dozens of confirmed events annually, contributing genuine, novel scientific insight into stellar evolution, the origin of heavy elements, the population and properties of black holes throughout the universe, and, as the next chapters of this book will explore, providing an entirely independent tool for measuring some of the same fundamental cosmological quantities, including the universe's expansion rate, that the supernova surveys and cosmic microwave background studies discussed earlier in this book have worked to pin down through very different observational means.

The scientific value of this rapidly growing catalog extends well beyond any single spectacular event like GW170817. With dozens, and eventually hundreds, of confirmed black hole mergers now available for study, astrophysicists can begin doing genuine population-level statistics: examining the distribution of black hole masses across the full catalog, searching for patterns that might reveal how these objects actually form and evolve throughout cosmic history. One particularly interesting puzzle to emerge from this accumulating catalog involves black holes with masses in a specific range, roughly fifty to one hundred and thirty times the mass of the Sun, a range where conventional stellar evolution theory predicts that ordinary massive stars should not be able to produce black holes directly, because stars in this mass range are expected to be entirely disrupted by a specific kind of thermonuclear explosion, called a pair instability supernova, that leaves no compact remnant behind at all. And yet gravitational wave observations have detected mergers involving black holes that fall squarely within this theoretically forbidden mass gap, a genuine puzzle that has revived serious scientific interest in alternative formation channels, including the possibility, discussed briefly in chapter four of this book, that at least some black holes in this mass range might be primordial black holes, formed not from the collapse of dying stars at all but from extreme density fluctuations in the first fraction of a second after the Big Bang itself, a hypothesis that remains actively debated but that gravitational wave astronomy, uniquely among all available observational techniques, is positioned to help resolve, simply by continuing to accumulate a larger and more statistically robust catalog of merger events across a wide range of masses.

The instrument Rainer Weiss first sketched on paper in 1972 has become something considerably larger than any single detection, however historic: a genuinely new sense organ for humanity, attuned not to light but to the trembling of spacetime itself, and it is still, as of this writing, remarkably early in what that new sense is likely to reveal.

There is a final point worth making before this chapter closes, one that connects gravitational wave astronomy back to the broader themes running through this entire book. Every merger detected since 2015 represents an independent data point about how black holes and neutron stars form, pair up, and eventually collide across the history of the universe, and taken together, this growing catalog has begun to function as its own kind of cosmological probe, entirely separate from the supernova surveys, cosmic microwave background studies, and galaxy clustering measurements discussed in earlier chapters. Because the strength of a gravitational wave signal depends on the distance to its source in a manner general relativity predicts precisely, while a neutron star merger's electromagnetic counterpart can independently reveal the recession velocity of its host galaxy through ordinary redshift measurements, events like GW170817 offer a genuinely new, independent method for measuring the Hubble constant discussed in the previous chapter, a technique cosmologists have named the standard siren method, echoing the standard candle terminology used for supernovae. As the catalog of neutron star mergers with confirmed electromagnetic counterparts grows in the coming years, particularly once the next-generation detectors discussed in chapter fifteen come online, this standard siren technique has the potential to provide an entirely independent check on the Hubble tension, one that depends on neither the Cepheid-based distance ladder nor the cosmic microwave background analysis, and that could, in principle, help determine which of those two existing methods, if either, is affected by an unaccounted source of systematic error. Gravitational wave astronomy, in other words, has grown from a single historic detection into something considerably more versatile: not merely a new way of confirming Einstein's century-old prediction, but a genuinely new observational tool with the potential to help resolve one of the most stubborn open questions in modern cosmology.

CHAPTER THIRTEEN
Weighing the Invisible

This book opened with a question worth returning to directly now that eight chapters have been spent circling around dark matter and dark energy from different angles: how, exactly, do you weigh something that cannot be seen? The honest answer is that physicists have developed not one method but several, each independent, each exploiting a different physical consequence of mass and energy, and it is the agreement between these independent methods, rather than any single measurement alone, that gives modern cosmology its confidence. This chapter steps back from the historical narrative of the preceding chapters to lay out, side by side, the actual toolkit astronomers use to measure things with no mass in any conventional sense, or with mass too dim to see directly, and to show how these tools reinforce and cross check one another.

The first and most direct tool, already introduced in earlier chapters, is gravitational lensing, a phenomenon predicted by general relativity and confirmed spectacularly during Eddington's 1919 eclipse expedition. Mass, whether visible or dark, bends the path of light passing near it, and this bending comes in two distinct flavors, each useful for a different purpose. Strong gravitational lensing occurs when an extremely massive, extremely concentrated object, typically a galaxy cluster, sits directly between Earth and a more distant background galaxy, bending that background galaxy's light so dramatically that it produces dramatic, unmistakable visual distortions: elongated arcs stretched around the lensing cluster, or in especially precise alignments, multiple duplicate images of the same background galaxy appearing at different points around the lens, an effect that looks, when photographed by a large telescope, almost like a funhouse mirror reflection stretched across the sky. Because the degree of this distortion depends predictably on the total mass of the lensing object, strong lensing provides astronomers with a direct, calculable measurement of a galaxy cluster's total mass, entirely independent of how much of that mass is visible starlight versus invisible dark matter.

Weak gravitational lensing operates on the same underlying physical principle but is far more subtle, and consequently, far more broadly useful. When a foreground mass is not concentrated and massive enough to produce dramatic, individually visible distortions in background galaxies, it nonetheless introduces a small, statistically detectable warping into the shapes of the many faint background galaxies visible behind it. No single background galaxy's distortion is individually meaningful, since galaxies naturally come in a range of shapes and orientations to begin with. But by statistically analyzing the shapes of thousands or millions of background galaxies across a given patch of sky, astronomers can detect a coherent, systematic pattern of distortion, a subtle tendency for background galaxies to appear very slightly elongated in a consistent direction, that reveals the presence and distribution of foreground mass, dark matter included, even when that mass produces no dramatic strong-lensing arcs at all. This technique has become one of the primary tools for mapping the large-scale distribution of dark matter across enormous volumes of the universe, and it forms the observational backbone of ambitious modern survey projects, including the Vera C. Rubin Observatory mentioned in chapter three, whose planned decade-long survey is expected to generate the most detailed weak lensing map of the universe's dark matter distribution ever assembled.

Nowhere has gravitational lensing produced more persuasive evidence for dark matter's physical reality than in the study of a particular pair of colliding galaxy clusters known, because of its distinctive appearance, as the Bullet Cluster, formally catalogued as 1E 0657-56. Located roughly 3.7 billion light years from Earth, the Bullet Cluster represents the still-unfolding aftermath of a collision between two separate galaxy clusters, a collision astronomers estimate occurred roughly 3.8 billion years ago and that remains, in a very real sense, still happening, its consequences still visible and measurable today. What makes this particular collision so scientifically valuable is that it separated two different components of each cluster's mass in a way that allows astronomers to study them independently, providing an almost laboratory-clean test of the dark matter hypothesis against its most serious rival explanation.

Here is how that separation worked. Each of the two colliding galaxy clusters contained three distinct components: the visible galaxies themselves, which are relatively small and compact and therefore passed through each other during the collision largely unaffected, much like two loosely packed swarms of gnats passing through one another; enormous clouds of hot, X-ray emitting intracluster gas, which actually constitutes the majority of each cluster's ordinary, visible matter and which, unlike the individual galaxies, does interact through ordinary electromagnetic and pressure forces, causing the two gas clouds to slow down dramatically and lag behind as the clusters passed through each other, similar to how two clouds of smoke would collide and become tangled rather than passing cleanly through one another; and, if dark matter theory is correct, invisible dark matter halos surrounding each cluster, which, like the visible galaxies, should interact with each other only through gravity and should therefore pass through the collision largely undisturbed, continuing onward essentially unimpeded, just as the visible galaxies did.

Using X-ray telescope observations, astronomers can directly map where the hot, slowed-down intracluster gas ended up after the collision, since it emits X-rays that space telescopes like NASA's Chandra X-ray Observatory can detect and precisely localize. Using weak gravitational lensing, astronomers can independently map where the majority of each cluster's total gravitational mass is actually located, regardless of whether that mass is visible or dark. When astronomers first performed this comparison for the Bullet Cluster, in a landmark 2006 study led by Douglas Clowe, the result was about as clean and dramatic as any single observation in modern astrophysics has ever managed to be: the two maps did not match. The X-ray gas, representing the majority of each cluster's ordinary, visible matter, sat concentrated near the collision's center, where it had collided, slowed, and piled up. But the gravitational lensing map, representing where the true bulk of each cluster's mass actually resided, showed two separate concentrations, offset from the gas and instead tracking closely with the positions of the visible galaxies, which had passed through the collision largely undisturbed.

This offset, measured with a statistical confidence exceeding eight standard deviations in subsequent, more detailed analyses, represents extraordinarily strong evidence against the primary rival explanation for the missing mass problem discussed back in chapter two: the idea that Newtonian gravity itself might simply behave differently at large scales, requiring no additional dark matter at all, a family of theories generally referred to as Modified Newtonian Dynamics, or MOND. A pure modified-gravity explanation predicts that a cluster's gravitational effects should track its visible, baryonic mass, which in the Bullet Cluster's case would mean the gravitational lensing signal should follow the X-ray gas, since that gas constitutes the overwhelming majority of each cluster's ordinary matter. Instead, the lensing signal clearly followed the collisionless, largely invisible component, exactly what the dark matter hypothesis predicts and precisely what a pure modified-gravity explanation, without invoking additional unseen mass, struggles to reproduce. Multiple attempts by researchers sympathetic to modified gravity theories to reconcile the Bullet Cluster's observations with MOND-style models, without invoking dark matter, have consistently run into serious quantitative difficulty, and the Bullet Cluster remains, nearly two decades after its initial analysis, one of the single most frequently cited pieces of observational evidence in the dark matter debate, precisely because it isolates and directly tests the specific behavior, mass tracking visible matter versus mass tracking a separate, collisionless component, that most cleanly distinguishes between the two competing explanations.

A second major tool in the astrophysicist's toolkit for weighing the invisible universe comes not from individual galaxies or clusters but from the universe as a whole, in its infancy: the cosmic microwave background, the faint, pervasive radiation left over from roughly 380,000 years after the Big Bang, when the universe had cooled enough for neutral hydrogen atoms to form for the first time, allowing light to travel freely through space rather than being constantly scattered by a hot, ionized plasma. This radiation, discovered accidentally in 1965 by the radio astronomers Arno Penzias and Robert Wilson, who initially mistook the persistent background signal for interference from pigeon droppings on their antenna before realizing they had detected genuine cosmic radiation, carries within it an extraordinarily detailed imprint of the early universe's composition and structure. Tiny temperature variations across the cosmic microwave background, mapped with exquisite precision by satellite missions including NASA's WMAP and the European Space Agency's Planck mission, encode the specific pattern of sound waves, called baryon acoustic oscillations, that rippled through the early universe's hot plasma before the moment light was set free. The precise size and spacing of these ripples depends sensitively on the relative proportions of ordinary matter, dark matter, and dark energy present in the early universe, allowing cosmologists to extract remarkably precise values for each component's contribution to the universe's total mass and energy budget, values that agree, to a striking degree, with the independent measurements derived from supernova surveys and gravitational lensing.

A third tool involves the large-scale distribution of galaxies themselves across the observable universe, mapped systematically by enormous galaxy survey projects, revealing a filamentary, web-like structure sometimes called the cosmic web, with galaxies clustering along vast filaments and sheets surrounding enormous, relatively empty voids.

Computer simulations of how this structure should form, run using the known laws of gravity and starting from the initial conditions revealed by the cosmic microwave background, only successfully reproduce the actual observed pattern of galaxy clustering when dark matter is included as a major component of the simulated universe. Simulations run without dark matter, using ordinary matter alone, consistently fail to produce anything resembling the observed cosmic web, since ordinary matter alone, interacting primarily through the relatively weak forces of gravity and radiation pressure in the early universe, simply does not clump together quickly enough, or in the right pattern, to produce the large-scale structure astronomers actually observe within the available time since the Big Bang. Dark matter, unaffected by radiation pressure and interacting only through gravity, was able to begin clumping together far earlier in cosmic history, providing the gravitational scaffolding that ordinary matter later fell into, eventually forming the galaxies, galaxy clusters, and filaments that populate the universe as we observe it today.

What should strike you, surveying these methods together, gravitational lensing at the scale of individual clusters, the cosmic microwave background from the early universe, and the large-scale structure of galaxies across cosmic history, is how genuinely independent they are of one another, relying on entirely different physical processes, different instruments, and different eras of cosmic history, and yet converging on remarkably consistent values for how much of the universe consists of ordinary matter, dark matter, and dark energy.

It is worth adding a bit more detail to the cosmic microwave background story, because the discovery itself is one of the more charming accidents in the history of science, and because the instrument involved illustrates just how much careful, patient engineering underlies even a discovery that arrived largely by chance. Penzias and Wilson, working at Bell Labs in New Jersey in 1964 and 1965, were not looking for evidence of the Big Bang at all. They were engineers tasked with calibrating an extremely sensitive microwave antenna, originally built for satellite communication research, and eliminating every possible source of interference from their measurements. No matter what they tried, including physically removing a pair of pigeons that had nested inside the antenna and cleaning out what they later described, with evident distaste, as a white dielectric material the pigeons had left behind, a persistent, uniform background hiss remained in their data, present regardless of which direction they pointed the antenna, which time of day they measured, or which season of the year it was. Unbeknownst to Penzias and Wilson, a team of theoretical physicists at nearby Princeton University, led by Robert Dicke, was simultaneously developing the theoretical prediction that the Big Bang should have left behind exactly this kind of faint, uniform microwave background radiation, and was actively building its own instrument to search for it. A phone call between the two groups, arranged after a colleague recognized the connection between Penzias and Wilson's stubborn interference signal and the Princeton group's theoretical prediction, led to a pair of companion papers published jointly in 1965, one reporting the observation, the other providing the theoretical interpretation, together representing one of the most significant confirmations of Big Bang cosmology in the theory's history. Penzias and Wilson received the Nobel Prize in Physics in 1978 for the discovery, a nice, if slightly bittersweet, illustration of how major scientific breakthroughs sometimes arrive not through a targeted search but through the diligent, careful work of researchers simply trying to eliminate every possible source of error from an unrelated engineering problem, until the only remaining explanation turns out to be genuinely new physics.

The specific pattern of temperature variations within the cosmic microwave background, the acoustic peaks that encode the universe's early composition, deserves a slightly fuller explanation as well, since the underlying physics rewards a closer look. In the hot, dense plasma filling the early universe, before neutral atoms could form, ordinary matter and radiation were tightly coupled together, behaving essentially as a single fluid, while dark matter, interacting only through gravity, was not coupled to the radiation in this way. Slight density variations present from the universe's earliest moments acted as gravitational wells, pulling the combined matter-radiation fluid inward, while the radiation pressure within that fluid simultaneously pushed outward, setting up genuine acoustic oscillations, sound waves, propagating through the early universe's plasma in a manner not unlike sound waves propagating through air, though on a cosmic scale almost impossible to intuitively grasp. When the universe cooled enough for neutral atoms to form and the plasma suddenly became transparent, these sound waves were effectively frozen in place, imprinted permanently onto the cosmic microwave background as a characteristic pattern of temperature variations at specific angular scales across the sky. Because dark matter's gravitational wells, unaffected by radiation pressure, deepened the acoustic oscillations in specific, calculable ways, while ordinary matter's coupling to radiation pressure damped them in a different, equally calculable way, the precise height and spacing of these acoustic peaks, when analyzed carefully, allow cosmologists to separately extract the density of ordinary matter and the density of dark matter present in the early universe, an extraordinarily elegant piece of physics that transforms what looks, to the untrained eye, like a fairly unremarkable, blotchy microwave map of the sky into one of the most information-rich data sets in all of modern cosmology.

Zwicky, working alone with a slide rule and a handful of galaxy velocities in 1933, could never have anticipated just how many independent lines of evidence would eventually confirm the essential correctness of his strange, dismissed idea. But that is precisely what happened, and it stands as one of the more satisfying vindications in the history of modern physics. This is, in the end, the methodological heart of modern cosmology: you weigh the invisible not through any single clever trick, but by approaching the same underlying question from as many genuinely independent directions as possible, and treating convergence between those independent approaches as the strongest form of evidence science has available.

CHAPTER FOURTEEN
When Worlds Don't Add Up

Every chapter of this book so far has told a version of the same reassuring story: physicists notice something strange, develop independent methods to investigate it, and those independent methods converge on a consistent answer, transforming an anomaly into an accepted piece of cosmological knowledge. This chapter tells a different kind of story, one without a tidy resolution, because the tension it describes remains genuinely unresolved as of this writing, and has, if anything, grown more persistent and more statistically significant with each passing year of increasingly precise measurement.

At the center of this chapter sits a single number: the Hubble constant, denoted H0, which describes the current rate at which the universe is expanding, expressed in units of kilometers per second per megaparsec, essentially telling you how much faster a galaxy appears to recede from us for every additional unit of distance separating it from Earth. This is, in principle, exactly the kind of number the methods described throughout this book should be able to pin down with precision, and in a sense, they can. The trouble is that two entirely different, independently rigorous methods for measuring it produce two different answers, and neither side of the disagreement has been able to identify a convincing error in the other's methodology, despite a full decade of intense, sustained scrutiny.

The first method, often called the local or late-universe method, measures the Hubble constant directly, using nearby astronomical objects whose distances astronomers can determine through a carefully calibrated sequence of overlapping techniques known as the cosmic distance ladder. This ladder typically begins with Cepheid variable stars, a class of star whose brightness pulsates in a regular, predictable pattern directly related to its true luminosity, a relationship discovered by the astronomer Henrietta Swan Leavitt in the early twentieth century and still foundational to distance measurement today. By comparing a Cepheid's known true luminosity against its observed apparent brightness, astronomers can calculate its distance, and by observing Cepheids in galaxies that also happen to host Type Ia supernovae (the same standard candles discussed in chapter six), researchers can calibrate the supernovae's true brightness against the more precisely known Cepheid distances, then use that calibration to measure the Hubble constant using a much larger sample of more distant supernovae, whose host galaxies are too far away for individual Cepheids to be resolved. The leading research effort using this approach, called SH0ES, led by Adam Riess (the same Riess whose earlier work on cosmic acceleration was discussed in chapter six), has consistently produced a local Hubble constant value clustering around 73 kilometers per second per megaparsec, with a 2024 measurement, combining Cepheid distances with an expanded sample of supernovae, arriving at 73.17 kilometers per second per megaparsec, and an even more recent 2025 community consensus estimate, combining several independent local distance methods, converging on 73.50 kilometers per second per megaparsec, with a remarkably tight precision of just over one percent.

The second method takes an entirely different approach, working backward from the early universe rather than forward from nearby galaxies. Using the detailed acoustic peak pattern within the cosmic microwave background, discussed in the previous chapter, cosmologists can determine the universe's precise composition, its density of ordinary matter, dark matter, and dark energy, at the moment that radiation was released, roughly 380,000 years after the Big Bang. Given that composition, and assuming the standard Lambda-CDM cosmological model correctly describes how the universe has evolved since that early moment, physicists can calculate, rather than directly measure, what the present-day expansion rate should be. The Planck satellite mission, whose detailed maps of the cosmic microwave background have provided the most precise such measurement to date, produces a Hubble constant value of 67.4 kilometers per second per megaparsec, a number confirmed and refined by subsequent analyses combining Planck's data with results from the Dark Energy Spectroscopic

Instrument survey discussed in chapter seven, which independently constrains the same underlying cosmological model using large-scale galaxy clustering data.

Seventy three versus sixty seven. On its face, this might not sound like an enormous discrepancy, roughly nine percent, but the significance of a discrepancy in physics is not measured by its raw size alone. It is measured against the precision of the measurements producing it, and this is where the Hubble tension becomes genuinely alarming to cosmologists rather than a minor rounding disagreement. Both measurements have become so precise, with uncertainty margins shrinking well below one or two percent on each side, that the gap between them now represents a statistical tension exceeding five standard deviations by most recent analyses, and in some more recent assessments incorporating the tightest available local measurements, exceeding seven standard deviations. In physics, a five-sigma threshold is conventionally treated as the standard for claiming a genuine discovery, meaning the probability of this discrepancy arising purely from random statistical chance, if both measurements were actually sampling the same true underlying value, is smaller than roughly one in three point five million. This is not the kind of gap that comfortably resolves itself as more data accumulates and error bars shrink toward overlap, the way many apparent tensions in physics eventually do. If anything, a decade of accumulating data and improving precision has made the disagreement between these two methods more statistically severe, not less, a pattern that tends to indicate something genuinely wrong, either in one of the measurement methods themselves or, more provocatively, in the standard cosmological model both methods ultimately depend on.

The obvious first suspect, whenever two independent measurements disagree this persistently, is systematic error: some subtle, unaccounted-for bias contaminating one measurement or the other. This possibility has received enormous scrutiny, and it is worth walking through some of the specific concerns that have been investigated and, in most cases, ruled out. Critics of the local, Cepheid-based measurement raised a reasonable concern that in distant, crowded galaxies, individual Cepheid stars might become blended together in telescope images with unrelated background stars, subtly and systematically biasing the measured brightness and therefore the calculated distance. This was a serious enough concern that it motivated dedicated follow-up observations using the James Webb Space Telescope, launched in 2021 with far sharper resolving power than its predecessor, the Hubble Space Telescope, specifically to check whether Cepheid crowding might explain the discrepancy. The results, reported through 2024 and 2025, found that James Webb's sharper observations confirmed rather than overturned the earlier Hubble Space Telescope Cepheid measurements, essentially ruling out crowding as the source of the tension, a genuinely significant result since it eliminated what had been considered one of the most plausible mundane explanations.

Researchers have also pursued entirely independent local distance measurement techniques that avoid Cepheids altogether, providing a valuable cross check on whether the tension might be an artifact specific to Cepheid-based calibration. These alternative approaches include using a different class of pulsating star called RR Lyrae variables, using the observed brightness and expansion properties of Type II supernovae, and using a technique called the tip of the red giant branch method, which exploits a distinct, predictable brightness threshold in the evolution of aging stars. Every one of these independent methods, despite using entirely different classes of astronomical object and entirely different calibration techniques, has continued to produce local Hubble constant values clustering in the low seventies, meaningfully higher than the cosmic microwave background prediction, a consistency across independent methods that makes a simple, single-source measurement error look increasingly unlikely as the explanation.

Not every research group agrees that the tension is fully established, and it is worth representing this genuine scientific disagreement honestly rather than presenting a false consensus. A research team led by the astronomer Wendy Freedman, using an independent calibration approach centered on red giant stars rather than Cepheids, has published results, including a notable 2025 status report using combined Hubble and James Webb Space Telescope data, that come out somewhat lower than the SH0ES team's Cepheid-based value, close enough to the Planck cosmic microwave background prediction that Freedman's team has argued the standard cosmological model may hold up without requiring new physics after all, once the local measurement's remaining systematic uncertainties are properly accounted for. This disagreement between two serious, methodologically careful research groups, both using legitimate but somewhat different calibration approaches, remains a genuine, actively contested point within the field, and it is a useful reminder that even well-established observational cosmology continues to involve real, unresolved methodological debate among working scientists, not a uniform, settled consensus waiting merely to be popularized.

If the tension does hold up, and the balance of current evidence across multiple independent local measurement techniques increasingly suggests that it will, the implications reach well beyond a mere technical disagreement over a single number. It would suggest that the standard Lambda-CDM cosmological model, remarkably successful as it has been at explaining an enormous range of other observations, as discussed throughout this book, may be missing some ingredient, some additional physics operating either in the early universe, altering how the cosmic microwave background prediction should be calculated, or in the nature of dark energy itself, perhaps varying over cosmic time in ways the simple constant cosmological constant model does not capture, a possibility already hinted at by the DESI survey results discussed in chapter seven. Proposed explanations range from modifications to the early universe's expansion history, involving new, currently unknown forms of energy active only in the universe's first few hundred thousand years, to more exotic possibilities involving new particle physics, to, more mundanely but still significantly, the possibility that our own galaxy happens to sit within an unusually large, underdense region of the universe, a cosmic void whose local gravitational effects could subtly bias local expansion rate measurements relative to the true cosmic average, a hypothesis actively being tested against new data as of this writing.

The Hubble tension is a useful chapter to include near the close of this book precisely because it resists the tidy, triumphant narrative arc that has characterized most of the discoveries described in earlier chapters. There is no equivalent here of Vera Rubin's flat rotation curves settling the dark matter question, or the Bullet Cluster's clean separation of mass components settling the modified gravity debate. The Hubble tension remains a live, unresolved puzzle, one that upcoming observational campaigns, including new results expected from the Euclid space telescope and continued refinement of both local and cosmic microwave background measurement techniques, are actively working to resolve in the coming years.

It is worth briefly mentioning that the Hubble tension is not the only unresolved discrepancy currently unsettling cosmologists, though it remains the most statistically severe and the most extensively studied. A second, related puzzle, sometimes called the S8 tension, involves a measure of how clumpy the universe's matter distribution appears to be at relatively recent cosmic times, derived from weak gravitational lensing surveys and galaxy clustering measurements discussed in the previous chapter. Several, though not all, weak lensing surveys have found the universe to appear slightly less clumpy than the standard Lambda-CDM model, calibrated using the cosmic microwave background, would predict, a milder tension than the Hubble constant discrepancy, typically reported in the range of two to three standard deviations rather than five or more, but one that some cosmologists suspect may ultimately be related to whatever new physics, if any, eventually resolves the more severe Hubble tension. Whether these two puzzles, the Hubble tension and the S8 tension, share a common underlying cause, or represent two entirely separate issues, possibly one genuine new physics and one lingering systematic measurement error, remains an open and actively debated question, illustrating that modern cosmology, for all its genuine successes, still has real, substantive homework left to complete.

There is also a broader, almost philosophical point worth drawing out from this chapter before moving to the book's final two chapters, because it bears on how you, as a reader, should weigh everything presented in the preceding thirteen chapters. Confidence in dark matter's existence, established through the multiple independent lines of gravitational evidence discussed in chapter thirteen, does not depend on the Hubble tension being resolved one way or another. Confidence in cosmic acceleration and dark energy's existence, established through the supernova, cosmic microwave background, and large-scale structure evidence discussed in chapters six and seven, likewise does not hinge on this particular open question. What the Hubble tension actually threatens is not the broad, well-established picture of what the universe contains, roughly five percent ordinary matter, twenty seven percent dark matter, sixty eight percent dark energy, but rather the precise details of the standard model describing how that content has evolved and interacted across the universe's full history. Science of this kind, genuinely secure about its major conclusions while genuinely uncertain about specific, actively researched details, is not a sign of a field in crisis. It is closer to a sign of a field working exactly as it should, having settled its largest questions decisively enough to notice, and take seriously, the smaller but genuinely puzzling discrepancies that remain. It stands as an honest reminder that the Lambda-CDM cosmological model, for all its genuine successes cataloged throughout this book, remains a model, a remarkably good one, but one that current data suggests may not yet be the final, complete word on how the universe actually works.

CHAPTER FIFTEEN

What We Still Don't Know

A book like this one could easily give the impression that the four mysteries at its center, dark matter, dark energy, gravitational waves, and the quantum vacuum, are each essentially solved problems, differing only in how interesting their solutions turned out to be. That impression would be wrong, and this chapter exists to correct it directly, gathering together the genuinely open questions scattered throughout the preceding fourteen chapters and looking honestly at where each field actually stands, and where it is headed, as of this writing.

Start with dark matter, the subject of chapters two through five and again in chapter thirteen. Despite ninety years of accumulating gravitational evidence, evidence now about as solid as anything in observational astrophysics, the fundamental question of what dark matter actually is at the particle level remains unanswered. The underground xenon detectors described in chapter five, LZ, XENONnT, and PandaX-4T, continue to push the boundaries of WIMP sensitivity, and as each new null result arrives, the remaining viable parameter space for the simplest WIMP models continues to shrink. The field's next major step, already underway, involves a proposed successor experiment called XLZD, intended to combine the accumulated engineering knowledge from LZ, XENONnT, and the related DARWIN design concept into a single next-generation xenon observatory large enough to push WIMP sensitivity right up against the neutrino fog boundary discussed in chapter five, the point at which solar neutrino background becomes an irreducible limit on how sensitive this particular detection strategy can become. If WIMPs are not found by the time that boundary is reached, the field will face a genuine crossroads, requiring either fundamentally different detection strategies or a broader reconsideration of which theoretical candidates deserve the most attention and resources going forward. Axion research, meanwhile, continues to mature along a largely separate track, with experiments steadily expanding the range of axion masses they can rule out, gradually working through the theoretically favored parameter space one carefully engineered magnetic cavity at a time. It remains entirely possible, and worth saying plainly, that dark matter could turn out to be something not yet seriously considered by current theory, a reminder that even a field as mature as dark matter research retains real room for a genuine conceptual surprise.

Dark energy, the subject of chapters six and seven, presents perhaps the deepest open question in all of contemporary physics: the cosmological constant problem, the roughly sixty to one hundred and twenty order of magnitude mismatch between the vacuum energy density predicted by quantum field theory and the actual, much smaller value inferred from cosmological observation. No proposed resolution to this problem has achieved anything close to consensus acceptance within the theoretical physics community, and it remains, by a wide margin, one of the most significant unsolved problems bridging quantum physics and gravity, a genuine embarrassment sitting beneath an otherwise remarkably successful cosmological framework. Compounding this theoretical puzzle, the observational question of whether dark energy's strength is truly constant over cosmic time, or varies subtly as some recent DESI survey results have tentatively hinted, remains actively unresolved, and upcoming years of continued DESI data collection, together with results expected from the Euclid space telescope, are likely to sharpen or resolve this particular question considerably before the end of this decade.

Gravitational wave astronomy, the subject of chapters nine through twelve, stands in a genuinely different position from the other topics in this book: rather than facing a foundational open question about whether the underlying phenomenon exists, the field is instead entering a period of explosive observational growth, with a clear, well-funded, and technically detailed roadmap for the coming decades. The current LIGO, Virgo, and KAGRA detector network continues to be incrementally upgraded, and by the mid-2030s, an entirely new generation of ground-based detectors is expected to come online: Cosmic Explorer, planned as a facility in the United States featuring a single interferometer with arms stretching forty kilometers, ten times longer than LIGO's current four-kilometer arms, and Einstein Telescope, a European design conceived as an underground triangular arrangement of six interferometers, each with ten-kilometer arms, built underground specifically to minimize the low-frequency seismic noise that limits current surface-based detectors. Together, these next-generation instruments are expected to deliver roughly a tenfold improvement in sensitivity over current detectors, translating into the ability to detect gravitational wave sources across a dramatically larger volume of the observable universe, extending gravitational wave astronomy's reach back toward the era of the very first stars, and increasing the annual detection rate from the current several dozen events per year toward an anticipated hundreds of thousands of mergers, transforming gravitational wave astronomy from an already thriving discipline into something approaching routine, high-volume survey science.

Beyond these ground-based instruments, the European Space Agency's Laser Interferometer Space Antenna, LISA, currently planned for launch sometime in the 2030s, will place a gravitational wave detector in space for the first time, consisting of three spacecraft separated by millions of kilometers, forming a giant triangular interferometer capable of detecting gravitational waves at much lower frequencies than any ground-based instrument can access, frequencies produced by different astrophysical sources entirely, including the mergers of the enormous supermassive black holes that sit at the centers of galaxies, events far too slow and low frequency for LIGO's four-kilometer arms to ever detect, but well within reach of LISA's vastly larger, space-based baseline.

The scale of ambition behind these next-generation instruments deserves a moment of its own appreciation, because it illustrates how thoroughly gravitational wave astronomy has shifted, in the space of one decade, from a speculative long-shot to a mainstream, richly funded pillar of observational astrophysics. Cosmic Explorer's proposed forty-kilometer arms would make it, by a considerable margin, the largest precision optical instrument ever constructed by humanity, requiring an entirely new scale of vacuum engineering, mirror fabrication, and seismic isolation beyond anything LIGO's current four-kilometer detectors demand. Einstein Telescope's underground, triangular design, meanwhile, reflects hard-won lessons about the persistent challenge of low-frequency seismic noise that has limited every surface-based interferometer built to date, a limitation that simply cannot be engineered away at the surface no mater how sophisticated the isolation system, but that burying the entire instrument deep underground, in the manner of the dark matter detectors discussed in chapter five, can substantially overcome. Taken together, current projections suggest that once this next generation of instruments comes fully online, likely sometime in the mid to late 2030s, gravitational wave astronomy could plausibly detect signals originating from the very first generation of stars and black holes to form after the Big Bang, pushing the observable frontier of this entirely new astronomical discipline back to a cosmic era light itself struggles to reveal directly, since the earliest stars and galaxies remain notoriously difficult to observe using conventional telescopes, however powerful.

Complementing this ground and space-based effort, a global network of radio astronomers has spent decades monitoring an array of rapidly rotating neutron stars called pulsars, watching for extremely subtle, correlated timing variations across the entire array that would indicate the presence of a background hum of gravitational waves permeating the universe, produced by the combined, overlapping signal of countless supermassive black hole mergers throughout cosmic history, too numerous and too overlapping to resolve as individual events. This pulsar timing array approach, coordinated internationally through collaborations including the NANOGrav project in North America, reported its first tentative evidence for exactly this kind of gravitational wave background in 2023, a genuinely exciting development that continues to be refined and strengthened with additional years of pulsar timing data.

The quantum vacuum, the subject of chapter eight, occupies a somewhat different position still, since the core phenomenon, vacuum energy's physical reality, is about as well established experimentally as anything in this book, confirmed to sub-percent precision through decades of refined Casimir force measurements. The open questions here are less about whether the phenomenon exists and more about its deeper theoretical connections: whether a fuller understanding of vacuum energy at the quantum level might eventually help resolve the cosmological constant problem plaguing dark energy research, and whether increasingly precise laboratory measurements of the Casimir force might eventually reveal subtle deviations hinting at entirely new physics operating at short range, additional spatial dimensions or modified gravitational behavior that current theory does not predict.

It is worth stepping back, at this point, to notice a genuine and pleasing symmetry across all four of these open frontiers. Each is being attacked using the same basic methodological strategy that has characterized every successful investigation described throughout this book: build increasingly sensitive instruments, gather increasingly precise data, and use the honest, sometimes uncomfortable disagreement between independent measurements, whether that disagreement takes the form of a null result narrowing a search, a persistent statistical tension like the Hubble tension discussed in the previous chapter, or a genuinely novel signal like the pulsar timing array's tentative gravitational wave background, as the primary engine driving the field forward. None of the four mysteries at the center of this book arrived at their current, well-established status through a single decisive experiment. Each required decades of patient, cumulative, occasionally frustrating work, performed by large teams of scientists willing to spend careers on problems that offered no guarantee of resolution within their working lifetimes, a theme that has recurred, in one form or another, in nearly every chapter of this book.

There is also something worth saying about the particular moment in scientific history this book has been written to capture. As of this writing, in 2026, all four of the frontiers surveyed in this chapter, dark matter detection, dark energy characterization, gravitational wave astronomy, and quantum vacuum research, sit at genuinely active, rapidly evolving stages, with major new instruments, results, and potentially field-altering discoveries plausible within the coming five to ten years. A reader returning to this book a decade from now may well find that some of the open questions catalogued in this chapter have been resolved, in either direction, while others, quite possibly ones that seem settled today, have opened into entirely new puzzles nobody currently anticipates. This is not a weakness of the account this book has given. It is, properly understood, the ordinary condition of any field of science actively engaged with the genuine frontier of human knowledge, and it is precisely what makes following these particular mysteries, in real time, as this book has attempted to do, so genuinely rewarding.

It is worth adding a bit more texture to the pulsar timing array story mentioned above, because it represents one of the more genuinely novel additions to the gravitational wave detection toolkit developed since LIGO's original 2015 breakthrough, and because it illustrates a broader, recurring theme in this book: entirely new detection strategies, developed independently and using completely different physical principles, tend to open up entirely new observational windows rather than simply replicating what earlier instruments could already see. Pulsars, the rapidly rotating neutron star remnants first discovered in 1967 by the astronomer Jocelyn Bell Burnell, sweep a beam of radio emission across space with a regularity that rivals the best atomic clocks, particularly a specific subclass called millisecond pulsars, which rotate hundreds of times per second with extraordinary long-term timing stability. By monitoring an array of dozens of these millisecond pulsars scattered across the galaxy for many years, and watching for a specific, correlated pattern of tiny timing deviations across the entire array, a pattern general relativity predicts should appear if a background of extremely low-frequency gravitational waves is washing across the galaxy, researchers can, in effect, turn the entire galaxy into a gravitational wave detector far larger than any instrument humans could physically construct on Earth or in space. The tentative 2023 evidence for this background, reported jointly by NANOGrav and several international partner collaborations conducting similar pulsar timing programs, is believed to originate primarily from the combined, overlapping gravitational wave signal of countless supermassive black hole binary systems merging throughout the universe's history, each individual signal far too faint and far too slow, with orbital periods measured in years rather than the fractions of a second LIGO detects, to resolve individually, but collectively producing a detectable statistical background hum, quite unlike anything LIGO's four-kilometer interferometers were ever designed to detect.

There is a broader lesson embedded in that pulsar timing story worth carrying forward into this book's final chapter. Each of the four detection strategies surveyed across these fifteen chapters, underground xenon detectors listening for a single atomic recoil, space telescopes measuring the brightness of exploding stars billions of light years away, laser interferometers measuring distance changes smaller than a proton, and a galaxy-spanning array of pulsars turned into an accidental gravitational wave antenna, represents a genuinely different way of listening to the universe, developed by different communities of researchers, using different instruments, sensitive to different frequencies, different scales, and different categories of astrophysical event entirely. No single one of these approaches, however brilliantly executed, could have revealed everything this book has described. It required all of them, developed largely in parallel, occasionally overlapping and cross checking one another, to build up the current picture of a universe far stranger, far more thoroughly interconnected, and far more actively, excitingly unfinished than any single observational technique could have revealed on its own.

The final chapter of this book steps back from the technical frontier altogether, to ask a different, more personal kind of question: why any of this should matter to you, a reader who is very likely not a professional physicist, going about an ordinary life largely untouched, in any obvious practical sense, by whether dark matter turns out to be a WIMP or an axion, or by whether dark energy proves constant or evolving. The answer, it turns out, has less to do with any practical application, though some genuinely exist, and more to do with something considerably older and, in its own way, considerably more valuable.

CHAPTER SIXTEEN

Why This Matters to You

You will likely never operate a laser interferometer. You will almost certainly never descend a mile underground to check on a tank of liquid xenon, and the odds that you will personally resolve the cosmological constant problem that has quietly embarrassed theoretical physics for the better part of a century are, statistically speaking, not favorable. None of the four mysteries this book has spent fifteen chapters examining will change what you have for dinner tonight, what your commute looks like tomorrow, or how much you pay for groceries next month. It is worth asking, honestly and without embarrassment, why any of this should matter to you at all.

Here is one answer, the most practical one available, and it is worth taking seriously before moving to the deeper case this final chapter actually wants to make. The technologies developed to chase these four invisible phenomena have, again and again, produced genuine, unglamorous, practical spillover into fields with nothing to do with cosmology. The precision timing techniques developed for pulsar astronomy and gravitational wave detection have influenced the atomic clock technology underlying the GPS system in your phone. The vacuum engineering and precision optics pioneered for LIGO's four-kilometer arms have advanced techniques now used in semiconductor manufacturing and other precision industries. The image sensor technology refined for astronomical surveys, including the kind of large-scale sky surveys discussed in earlier chapters, has fed directly into the digital camera sensors in the phone in your pocket. Basic research of the kind this book describes, research undertaken with no promise of practical application, has a genuinely strong historical track record of eventually producing exactly that, though almost never in ways anyone could have predicted at the outset. This is a real and defensible reason to care about funding this kind of science, and it deserves to be stated plainly rather than treated as an afterthought.

But it is not, if this book is being honest with you, the actual reason its author wanted to write it, or the reason worth dwelling on as this book closes. The deeper case is older, and it has less to do with usefulness than with something closer to a basic feature of what it means to be a curious animal capable of asking questions about its own circumstances.

Consider, one final time, the scale of what this book has actually described. Roughly five percent of the universe's total content is the kind of matter you, this book, and everything you have ever directly touched or seen is made of. The remaining ninety five percent, dark matter and dark energy combined, is something else, something that ninety years of dedicated, careful, occasionally brilliant scientific work has managed to characterize gravitationally with real precision while still, as of this writing, not fully understanding what it fundamentally is. You are reading these words, right now, while embedded in a universe overwhelmingly composed of something science cannot yet name. That is not a comfortable fact, particularly. It does not flatter human self-importance the way earlier cosmological pictures, placing Earth at the center of creation, once did. But there is a specific, durable kind of satisfaction available in facing that fact clearly rather than looking away from it, the same satisfaction a careful reader takes in a well-constructed mystery novel that refuses to hand over an easy, tidy resolution.

It is worth pausing on that comparison for a moment longer, because it is not merely a rhetorical flourish. A well-constructed mystery novel withholds its resolution not out of cruelty but because the withholding is where the actual pleasure of the genre lives, in the careful accumulation of clues, the false leads chased down and eliminated, the moment when two seemingly unrelated details suddenly click into alignment. Every chapter of this book has, in its own way, followed exactly that structure, because that is genuinely how these particular discoveries unfolded in real historical time, not as tidy textbook summaries handed down from on high, but as long, uncertain investigations, pursued by specific people, over specific years, frequently unsure whether the puzzle they were chasing would ever actually resolve. Reading popular science that flattens this uncertainty into a smooth, inevitable-sounding narrative does readers a genuine disservice, because it teaches an entirely false lesson about what scientific knowledge actually feels like while it is being made. This book has tried, throughout, to preserve some of that genuine uncertainty, the doubt, the false starts, the decades of null results, precisely because that uncertainty is not a flaw in the story. It is the story.

Consider, too, the specific character of the people who did the work this book describes. Fritz Zwicky, difficult and unpopular, refusing to explain away an inconvenient number because refusing was professionally comfortable. Vera Rubin, kept out of the field's inner circles for a decade by outright discrimination, building a career anyway and eventually forcing an entire discipline to take a forty-year-old anomaly seriously through nothing more than the sheer weight of careful, repeated observation. Saul Perlmutter and Brian Schmidt, leading rival teams who each expected to measure deceleration and instead found themselves, independently, staring at evidence for something considerably stranger, having the intellectual honesty to report exactly what their data showed rather than what they had expected to find. Rainer Weiss, spending four decades of a working scientific life on an instrument that might, for entirely legitimate physical reasons, have simply never succeeded, and being honest, throughout that entire span, about not knowing whether it would. None of these people set out to become symbols of anything.

They were simply, each in their own particular way, unwilling to look away from a puzzle once it had genuinely caught their attention, and willing to do the unglamorous, patient work required to actually resolve it, rather than settling for an elegant guess.

That willingness, more than any specific fact about WIMPs or cosmological constants or interferometer arm lengths, is the actual subject this book has been circling for sixteen chapters. Curiosity, sustained not for a day or a week but across entire careers, applied not to questions with obvious practical stakes but to questions whose only real reward was understanding itself, more clearly, more precisely, more honestly than anyone had managed before. This is, when you strip away the specific vocabulary of xenon detectors and supernova standard candles, one of the oldest and most distinctly human activities there is: looking at something strange, refusing to accept a comfortable non-answer, and doing the patient work required to actually find out.

There is a particular kind of humility built into everything this book has described, and it is worth naming directly as this final chapter closes, because it cuts against a common misconception about what scientific confidence actually feels like from the inside. The physicists in this book were not confident because they had everything figured out. They were confident, when they were confident at all, because they had built methods rigorous enough to tell them clearly when they were wrong, and had used those methods honestly, following the results wherever they actually led, including toward conclusions, an accelerating universe, a mismatch between predicted and observed vacuum energy of one hundred and twenty orders of magnitude, a persistent five-sigma disagreement in the Hubble constant, that nobody had anticipated and that many would have preferred not to find. That particular combination, genuine intellectual humility paired with genuine methodological rigor, is not a compromise between confidence and doubt. It is, this book would argue, the actual engine of everything worthwhile that science has ever produced, and it is available, in smaller and larger ways, to anyone willing to look closely at a question rather than settle for the first comfortable answer that presents itself.

It is worth reflecting, too, on how differently each of the four mysteries in this book was actually discovered, because the variety itself carries a lesson. Dark matter emerged from a difficult, disagreeable Swiss astronomer refusing to dismiss an inconvenient calculation, and then, four decades later, from a woman who had been kept out of the field's inner circles refusing to stop looking at a puzzling printout. Dark energy emerged from two competing teams, driven partly by professional rivalry, who each expected one result and instead found themselves converging, independently, on something else entirely, and who trusted their own careful methodology enough to report it despite how strange it seemed. Gravitational waves emerged from a young physicist tasked with teaching a course he only shakily understood, working through the problem carefully enough to stumble onto an idea that would occupy the working lives of hundreds of scientists for the next four decades. And the quantum vacuum emerged from an industrial researcher investigating colloidal chemistry for a consumer electronics company, following an unexpected mathematical thread to a conclusion nobody in that laboratory had gone looking for. None of these discoveries arrived through a single brilliant insight isolated from all context. Each arrived through a particular person's stubborn attention to something that did not quite fit, combined with the accumulated methodological toolkit an entire scientific tradition had built up specifically to distinguish a genuine anomaly from an error. That combination, individual curiosity meeting collective, hard-won method, recurs so consistently across these sixteen chapters that it starts to look less like coincidence and more like the actual mechanism by which this kind of understanding gets built, one difficult, inconvenient observation at a time.

There is also something worth saying, briefly and honestly, about the limits of what this book has been able to offer. Sixteen chapters and roughly two hundred pages is enough space to walk through the major discoveries, the key figures, and the current experimental frontier of four genuinely enormous fields of physics, but it is not enough space to resolve genuine, ongoing scientific disagreement, nor should it try to. Where this book has described open questions, the nature of dark matter, the constancy of dark energy, the resolution of the Hubble tension, it has tried to represent honestly what is actually settled, what remains genuinely contested among working physicists, and where the balance of current evidence points, without pretending to a certainty the field itself does not yet have. That honesty is, in its own modest way, part of the same tradition this book has spent its pages describing: better to report the data as it actually stands, however unresolved, than to manufacture a false sense of closure for the sake of a tidier story.

You began this book, if the introduction did its job, wondering how anyone could possibly weigh something with no mass, measure something that bends reality without ever quite showing itself directly. You now know, with real specificity, how that has actually been done: through the careful, repeated, cross-checked convergence of independent methods, through instruments built at the very edge of what engineering allows, through decades of patient work performed by scientists willing to spend entire careers on questions that offered no guarantee of resolution. The shadow this book's title refers to, the invisible mass, the invisible energy, the invisible ripples in spacetime, the invisible restlessness of empty space itself, has never stopped weighing on the visible universe, shaping every galaxy, every collision, every expanding stretch of cosmic distance you could ever hope to observe. It has simply been waiting, patiently, for someone curious enough, and stubborn enough, to notice.

There is one last thought worth leaving you with, distinct from everything argued above, because it is less an argument than an invitation. Every discovery recounted in this book began with someone noticing that a number did not add up, that a signal looked stranger than it should, that a prediction and an observation refused to match no matter how carefully they were checked. You do not need a telescope or an underground detector to practice that same habit of attention in your own life. The same disposition that led Zwicky to trust an inconvenient calculation, that led Rubin to keep measuring rotation curves long after the pattern was already clear, that led two rival supernova teams to report a result neither had expected, is available to anyone willing to notice when something does not fit and to stay with that discomfort long enough to find out why, rather than reaching immediately for the most convenient available explanation. That habit, more than any specific fact this book has conveyed about WIMPs or cosmological constants or interferometer sensitivity, is the most durable thing worth taking from these sixteen chapters.

That noticing is not finished. As this book has tried to make clear throughout its final chapters, every one of the four mysteries at its center remains, in some genuine and specific sense, an open question, actively being pursued by real scientists, using real instruments, right now, as you read this sentence. There is no better time than the present to have paid attention to how these mysteries actually work, because the coming decade promises to bring some of them, quite possibly, to genuine resolution, transforming several of the open questions in the previous chapter into settled chapters of their own in some future edition of a book very much like this one. Whether that resolution arrives through an underground detector finally catching its first dark matter particle, through a space-based interferometer picking up the gravitational hum of the earliest black holes, or through some discovery nobody currently anticipates, entirely upending the careful picture this book has assembled, the invisible universe is not finished revealing itself. It is, if the history recounted across these sixteen chapters is any guide, only getting started.

FURTHER READING AND RESOURCES

Readers interested in exploring the topics covered in this book more deeply may find the following real, verifiable resources useful.

On dark matter and its detection, the LZ (LUX-ZEPLIN) collaboration publishes its results and public updates through the Sanford Underground Research Facility and Lawrence Berkeley National Laboratory. The XENON collaboration publishes similar updates through its home institution network coordinated via Columbia University and other partner institutions. The Vera C. Rubin Observatory, based in Chile and operated jointly by the National Science Foundation and the Department of Energy, publishes public updates on its Legacy Survey of Space and Time at rubinobservatory.org.

On dark energy and cosmology, the Dark Energy Spectroscopic Instrument collaboration, based at Kitt Peak National Observatory in Arizona, publishes its survey results through Lawrence Berkeley National Laboratory. NASA and the European Space Agency jointly maintain public archives of results from the WMAP and Planck cosmic microwave background missions.

On gravitational waves, the LIGO Scientific Collaboration maintains a public education and outreach portal at ligo.caltech.edu, including plain-language explanations of major discoveries and updates on Advanced LIGO's ongoing observing runs. The European Virgo collaboration and the Japanese KAGRA collaboration maintain similar public resources. The Nobel Prize organization maintains public, freely accessible summaries of the 2011, 2017, and 1993 physics prizes discussed in this book at nobelprize.org.

For readers wanting a deeper technical grounding, Kip Thorne's book Black Holes and Time Warps offers an accessible but rigorous account of general relativity's development written by one of LIGO's own founders. Readers interested in Vera Rubin's own account of her career may seek out her essay collection Bright Galaxies, Dark Matters.

For general astronomical data verification, NASA's Astrophysics Data System, freely accessible at ui.adsabs.harvard.edu, provides public access to the peer-reviewed scientific literature underlying most of the findings discussed throughout this book.

This book was written to reflect the state of scientific understanding current as of early 2026. Given the genuinely active pace of research described throughout its final chapters, readers are encouraged to consult these primary sources directly for the most current results.

www.ingramcontent.com/pod-product-compliance
Lightning Source LLC
LaVergne TN
LVHW100143061126
851857LV00010B/221